(ASE SCIENCE PRACTICE)

teaching secondary
BIOLOGY
SECOND EDITION

EDITOR: MICHAEL REISS

HODDER
EDUCATION
AN HACHETTE UK COMPANY

Biblio

• •

Titles in this series:

Teaching Secondary Biology	978 1444 124316
Teaching Secondary Chemistry	978 1444 124323
Teaching Secondary Physics	978 1444 124309

The Publishers would like to thank the following for permission to reproduce copyright material:
Photo credits p.132 Imagestate Media (John Foxx), **p.133** Beth Van Trees – Fotolia.com
Acknowledgements
Every effort has been made to trace all copyright holders, but if any have been inadvertently overlooked the Publishers will be pleased to make the necessary arrangements at the first opportunity.

Although every effort has been made to ensure that website addresses are correct at time of going to press, Hodder Education cannot be held responsible for the content of any website mentioned in this book. It is sometimes possible to find a relocated web page by typing in the address of the home page for a website in the URL window of your browser.

Hachette UK's policy is to use papers that are natural, renewable and recyclable products and made from wood grown in sustainable forests. The logging and manufacturing processes are expected to conform to the environmental regulations of the country of origin.

Orders: please contact Bookpoint Ltd, 130 Milton Park, Abingdon, Oxon OX14 4SB. Telephone: (44) 01235 827720. Fax: (44) 01235 400454. Lines are open 9.00 – 5.00, Monday to Saturday, with a 24-hour message answering service. Visit our website at www.hoddereducation.co.uk

© Association for Science Education 2011
First published in 2011 by
Hodder Education,
An Hachette UK Company
338 Euston Road
London NW1 3BH

Impression number	5
Year	2015

Cover photo © MASSIMO BREGA EURELIOS/SCIENCE PHOTO LIBRARY
Illustrations by Tony Jones/Art Construction
Typeset in ITC Galliard by Pantek Media, Maidstone, Kent
Printed and bound by CPI Group (UK) Ltd, Croydon, CR0 4YY
A catalogue record for this title is available from the British Library

ISBN: 978 1444 124316

Contents

Contributors

Susan Barker has worked in Science teacher education since 1990, firstly at the University of Warwick and now the University of Alberta, Canada, where she is Chair of Secondary Education. She has worked extensively with the British Ecological Society and Ecological Society of America within the context of ecological education. Her ideal classroom scenario is where students are outdoors in the field, in groups collecting data to help answer questions they have posed themselves based on observations of the world around them.

Mike Cassidy has taught Biology in school, college and university; initially in the North East of England, latterly in the Midlands. He has been a curriculum consultant for QCA and has co-ordinated PGCE Science courses in Nottingham and at Warwick. Until recently, Mike was Assistant Professor in Science Education at Warwick University and he is currently Visiting Lecturer at Durham University.

Ann Fullick has many years experience as a Biology teacher, Head of Science and A-level examiner. She writes biology text-books for secondary schools in the UK and in many countries around the world, and also develops web-based resources to support the science curriculum.

Jennifer Harrison has recently retired from her academic post at the University of Leicester School of Education, where she co-ordinated the science PGCE courses. She has researched and written about biology and science teaching and the mentoring and practice of new teachers in particular. Her science teaching interests have focused on issues relating to health and sex education with the aim of encouraging a wider recognition of the role of the science teacher in the personal development of young people.

Neil Ingram is a senior lecturer in science education in the University of Bristol, where he co-ordinates the PGCE Science course. He has written extensively for Nuffield curriculum projects and co-authored with Michael Roberts a GCSE biology textbook. He was formerly Head of Science at Clifton College, Bristol.

Jenny Lewis is a senior lecturer in biology education at the University of Leeds. In addition to initial teacher training she undertakes classroom-based research on teaching and learning and has a particular interest in genetics. She has written articles for a range of journals and worked collaboratively with teachers to produce a number of interactive teaching sequences, including one for genetics. Before working at the University of Leeds she spent eight years in the classroom.

Roger Lock is a lecturer in science education at the School of Education in the University of Birmingham. After a short spell in industry as a sales representative, he taught in three schools, ending up as Head of Science at Trinity School, Leamington Spa before moving into initial teacher training. Roger trained teachers and carried out research in the universities of Leeds and Oxford prior to taking up his current post in the late 1980s. He writes and researches issues linked to practical work, the use of living things in science lessons and the initial training and early careers of science teachers.

Neil Millar is head of biology at Heckmondwike Grammar School. He has written articles for the School Science Review and has contributed chapters for A-level biology textbooks. He wrote the 'Merlin' software package for biological statistics and has trained teachers in statistics for biology. Before teaching, Neil spent 13 years carrying out medical research in the UK and USA.

Michael Reiss is Pro-Director and Professor of Science Education at the Institute of Education, University of London, Chief Executive of Science Learning Centre London, Vice President and Honorary Fellow of the British Science Association, Honorary Visiting Professor at the universities of Birmingham and York, Director of the Salters-Nuffield Advanced Biology Project and an Academician of the Academy of Social Sciences.

Nigel Skinner taught Science and Biology at state secondary schools in Wiltshire for 10 years before moving into teacher education. He is currently a senior lecturer and PGCE Secondary Science course leader at the University of Exeter where he also carries out research into science teacher education and development.

After completing his PhD **David Slingsby** worked for 29 years as a classroom teacher, 22 of these as Head of Biology. He has remained a practising scientist through a long-term project on a Shetland National Nature Reserve which he has monitored for over 40 years. He has a long history of commitment to ecological education through the British Ecological Society, of which he served as Chair of Education from 1999 to 2008 and received the Award for service to the Society in 2006. He served as GCSE coursework Moderator in Biology and in Science 1 and was involved in the Salters-Nuffield Advanced Biology project as a member of the writing team and as Principal and Chief examiner. He is currently an Open University tutor, a consultant and writer on biological education, a Principal examiner in Cambridge PreU Biology and Editor of the *Journal of Biological Education*. He was made a Fellow of the Society of Biology in 2010.

Mark Winterbottom is a lecturer in science education at the Faculty of Education, University of Cambridge. He teaches on the Science/Biology secondary PGCE course and on the science MEd course. His current research interests are in teacher education, biology education, and the use of new technologies in teaching and learning. Prior to joining the Faculty, Mark taught science in upper schools for 5 years. Mark has written a variety of school textbooks and books for teachers, and has contributed to a book on research methods for PGCE and MEd students.

Acknowledgements

The authors and editor are very grateful to the following for their advice during the preparation of this book:

BIOTUTOR Discussion Group
Chris Brown
Sarah O'Connell
Lesley Paris
Robin Sutton
Stephen Tomkins
Les Weeks

■ Risk assessment

As a service to users, a risk assessment for this text has been carried out by CLEAPSS and is available on request to the Publishers. However, the Publishers accept no legal responsibility on any issue arising from this risk assessment; whilst every effort has been made to check the instructions for practical work in this book, it is still the duty and legal obligation of schools to carry out their own risk assessment.

■ Dedication

To Angela Hall and in memory of John Cheverton

Introduction

Michael Reiss

This book is one of a series of three ASE handbooks, the others being parallel volumes on chemistry and physics. It adopts a pragmatic yet enthusiastic approach to the teaching of biology to 11–16 year olds. The author team has kept in mind a teacher confronted with the task of teaching a specific topic, e.g. respiration or ecosystems, in the near future. What does such a teacher need to produce a series of effective lessons? The first edition of this book was published in 1999, 12 years ago. This second edition retains the structure of the first edition but includes a number of new authors and all chapters have been substantially revised and brought up to date.

■ Who is this book for?

In writing their chapters, authors have identified a range of likely readers:

- new or less experienced biology teachers – though almost every biology teacher should find much of value
- chemists, physicists and science generalists who find themselves teaching parts of the biology curriculum
- student teachers and their tutors/mentors
- heads of department who need a resource to which to direct their colleagues.

While we have taken into account the current UK syllabus requirements, we have not stuck closely to any one curriculum. We expect that this book will be appropriate to secondary biology teachers in every country.

■ What should you find in this book?

We expect you, the reader, to find:

- high quality, sensible and stimulating ideas for teaching biology to 11–16 year olds
- suggestions for extending the range of approaches and strategies that you can use in your teaching
- things to which students respond well; things that fascinate them
- confirmation that a lot of what you are already doing is fine. There often isn't a single correct way of doing things.

■ How can you find what you want?

After careful discussion, the author team for the first edition divided secondary biology up into ten areas. We have retained this arrangement for this edition and these ten areas correspond to Chapters 1 to 10. It should be clear from the Contents page (page iii) what each chapter roughly consists of. A more detailed indication is given by the content boxes at the beginning of each chapter. These content boxes divide each chapter up into relevant sections. For specific topics, consult the index to the whole book. This contains the terms you might want to look up; e.g. 'blood', 'heart', 'circulation', 'blood vessels' and 'oxygen' are included rather than the name of every possible component of the circulatory system.

■ What is in each chapter?

Each chapter contains:

- content boxes, which divide up the chapter into shorter sections
- a set of possible teaching routes through the chapter
- brief coverage on what students may have learnt or experienced about the topic in their primary science lessons
- an outline teaching sequence showing how concepts can be developed throughout the 11–16 phase
- warnings about pitfalls
- information about likely student misconceptions
- helpful information about practical work and apparatus, e.g. how to prevent things from going wrong

- issues to do with safety (highlighted by the use of an icon in the margin)
- suggestions for the use of ICT
- suggestions about the use of books, the internet (shown by the arrow icons in the margin) and other resources
- opportunities for the teaching of applied or ethical aspects of biology
- opportunities for investigative work and other aspects of 'how science works'
- links with other areas of biology or science.

Finally, if you have any comments that you would like to make about this book please feel able to send them to me:

Michael Reiss
Institute of Education, University of London
20 Bedford Way, London WC1H 0AL
e-mail: m.reiss@ioe.ac.uk

Cells and life processes

Nigel Skinner

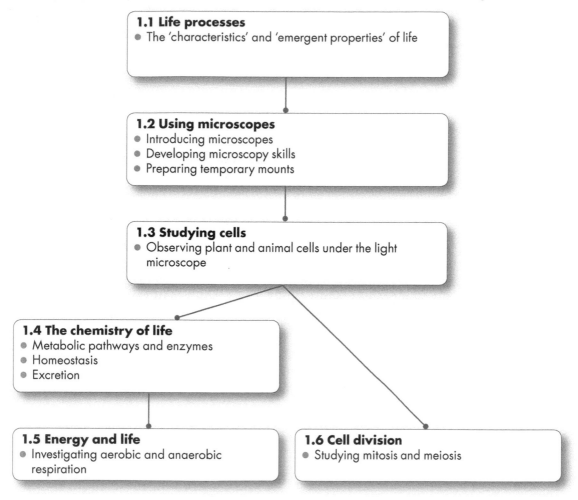

1.1 Life processes
- The 'characteristics' and 'emergent properties' of life

1.2 Using microscopes
- Introducing microscopes
- Developing microscopy skills
- Preparing temporary mounts

1.3 Studying cells
- Observing plant and animal cells under the light microscope

1.4 The chemistry of life
- Metabolic pathways and enzymes
- Homeostasis
- Excretion

1.5 Energy and life
- Investigating aerobic and anaerobic respiration

1.6 Cell division
- Studying mitosis and meiosis

Choosing a route

Life can be thought of as the result of the various interactions between the many different chemical substances that make up an organism. Biology involves the study of life at various levels of organisation. In ascending order of size and complexity these levels are:

molecular → cellular → tissue → organ
organ system → whole organism → population
community → ecosystem → biosphere

Studying biology at the molecular level involves dismantling organisms into their chemical components to investigate the structure and behaviour of the molecules that form them. At the cellular level biology involves finding out how these 'molecules of life' are organised and the ways in which they interact with each other. Defining what we mean by 'life' is not easy but it is at this level, in the transition from molecules to cells, that we begin to recognise the things that differentiate life from the absence of life. This chapter begins with a discussion of some of the characteristics of life that are displayed by whole organisms. The remainder of the chapter is mainly concerned with life processes at the molecular and cellular level. An understanding of these underpins the study of the many other aspects of biological science that are addressed in other chapters.

The activities suggested in the first three sections are aimed in the main at students aged between 11 and 14 years old and are best taught in the sequence in which they are presented. 'Life processes' outlines some of the differences between living and non-living things and the suggested approach is intended to help students gain an overview of what biology is all about. Microscopes are essential tools in biology and, in the second section, activities that could be used to familiarise students with their use are outlined. A topic that clearly requires the use of microscopes is the study of cells, and the third section includes approaches to cell biology suitable for both younger and older students.

The topics covered in the remaining sections are conceptually more demanding and are intended for use with older students. They can be taught as separate topics or could be integrated into larger topics. For example, study of the molecules found in living organisms is often initially addressed in the context of human diet and digestion, respiration is often first taught in the context of gas exchange, and cell division in connection with growth and reproduction. Whatever approach is used when teaching these topics it is very important to emphasise the links that students need to make with related topics to help them achieve a deeper understanding of the nature of life and living things.

1.1 Life processes

In introducing biology as the study of life and living processes, teachers have for many years taught their students about the seven 'characteristics of life' – **m**ovement, **r**eproduction, **s**ensitivity, **g**rowth, **r**espiration, **e**xcretion and **n**utrition – leading to the familiar mnemonic 'Mrs Gren'.

Some aspects of these are relatively easy for students to understand but it is not easy to convince students that these characteristics are common to all living things. For example, persuading students that plants can move and need oxygen for respiration can be problematic. Textbooks published in the USA often refer to the 'emergent properties of living things' (see Table 1.1) and you may find it useful to refer to these when discussing the differences between living and non-living things with students.

Table 1.1 The emergent properties of living things

Living things:

- are highly organised and more complex in structure than non-living things
- take energy from their environment and transform it from one form to another
- regulate their 'internal environment' (the technical term for this is homeostasis)
- respond to stimuli
- grow, develop and reproduce
- through evolutionary pressures become adapted to their mode of life
- contain within themselves the information by which they carry out the other functions characteristic of living things.

A key idea that needs emphasising is that in multicellular organisms, life processes are supported by the organisation of cells into tissues, organs and organ systems. Subsequent chapters will focus on the ways in which particular organs and organ systems support each of the life processes.

Previous knowledge and experience

Students will know about many life processes, but their understanding of the difference between living and non-living things tends to be very human- or animal-centred. For example, younger children sometimes think that if something moves or makes noises then it must be alive. Clearly, many living things can do neither and many non-living things can do both. Many of the words we use when talking about different types of living thing have a different meaning in everyday language from their scientific meaning. The word 'animal' is often used by children when referring only to the living things that have four legs, are furry and live on land. In a biological classification these are all terrestrial mammals, a subset of the whole animal kingdom, which includes humans (a concept that many children find difficult). Similarly, the word 'plant' is often used in a very restricted way to mean only those types of plant grown in 'plant pots' or the smaller plants ('flowers' and 'weeds') found in gardens. Mosses, grasses, trees and shrubs are often not considered to be plants by children, although to a biologist they clearly are.

A teaching sequence

When teaching 11 to 14-year-old students about life processes it is not appropriate to include detailed discussion of respiration or excretion. It is better to focus on the life processes of reproduction, growth, nutrition, movement and responsiveness, and to stress that all living things display these processes although in some cases they may not show them in an obvious way.

■ Recognising life

To introduce the idea that all living things display certain characteristics, groups of students could examine a range of easily obtainable organisms and discuss the features they have that enable them to recognise them as living. The organisms chosen for this study can include some that students will regard as obviously living and some, such as lichens, yeast and dormant plant seeds, that students may not at first consider to be living.

The 'sensitive plant' (*Mimosa pudica*) is a useful specimen to include because when its leaves are touched they rapidly droop downwards, illustrating very clearly that this plant can respond to stimuli and also move. Discussion of why organisms move can be linked with the process of nutrition. Animals need to move to gather food and to prevent themselves becoming food for another animal. Plants, providing that they are growing in suitable conditions, have all the things they need for photosynthesis and growth provided by their immediate environment but do show growth movements (tropisms) as described in Chapter 5. Animals that do not move from place to place, e.g. many marine organisms such as adult barnacles and mussels, live in an environment that is moving – their food comes to them and they filter it out of the water around them. Students could also compare the characteristics of living organisms with some non-living things that do things that might initially suggest that they are living, e.g. battery-powered toys or models that move and appear to be responsive – the key living process that they do not display is reproduction.

■ Studying life cycles

Many living processes can be discussed in connection with the sequence of events that take place from the start of life in a new individual through to the stage when the new individual is itself capable of reproduction. Animals such as stick insects are easily maintained in school laboratories and students can study their growth and development from the egg to the adult stage. This could include a consideration of their food preferences and patterns of movement.

Plants that complete their life cycle in a relatively short space of time can also be studied. For example, pea or bean seeds planted in early spring will become mature and produce seeds before the end of the summer term. 'Fast plants' (more correctly called rapid-cycling Brassicas) complete their life cycle in about 35 days if kept in continuous light. Because they require little space students can grow their own individual plants and these can be used to study a range of different life processes including growth, development, plant responses and nutrition. Details of how to grow and use these plants in schools can be obtained from Science and Plants for Schools (SAPS) whose website address is given at the end of this chapter.

Further activities

- Living things that do not show any obvious signs of life, e.g. frog-spawn, seeds and lichens, could be brought into lessons and students asked to explain how they would try to find out whether they are actually alive.

Enhancement ideas

- Students are often intrigued by the possibility that life exists on other planets and they could be asked to devise a series of investigations that could be carried out to find out whether life exists elsewhere in the Universe.
- Students could draw and label an imaginary animal or plant that displays the characteristics of life and is adapted to living in particular conditions, e.g. very hot or very cold climates. You could then get the students to swap their drawings with each other and draw potential predators or prey for the imaginary animals or plants. Their ideas could be displayed as posters.

1.2 Using microscopes

Microscopes are essential tools in biology and, if used appropriately, can play a vital role in developing students' knowledge and understanding of many aspects of living processes and organisms. In order to capitalise on the learning opportunities that microscopes offer it is important for students to be motivated by the experience of using a microscope. To achieve this they should be taught how to use a microscope correctly and be provided with material to study that will instil in them a sense of wonder at the intricacies of the microscopic world.

Most schools have sets of optical microscopes and the following sections provide details of how to introduce and use these with students. Many schools now have digital or USB microscopes which are relatively easy to use and with the appropriate software enable images to be captured and annotated on computers and used in conjunction with interactive whiteboards. USB microscopes use incident rather than transmitted light so are good for looking at surface features of specimens and will also provide good images of slides placed on a white background.

Previous knowledge and experience

All students will have seen magnified images in books, magazines and on television. Many will have used magnifying glasses to study small things and some will have used (and may even own) a simple microscope. Very few students will have received systematic instruction in the use of microscopes.

A teaching sequence

■ The importance of microscopes in biology

Before using microscopes it is worth showing students some images (using overhead transparencies, video, CD-ROMs or images from websites) that have been gained with light microscopes or scanning electron microscopes. Start lessons by displaying high-quality images that are likely to engender interest and discuss how these images were obtained and the value of being able to magnify things when trying to find out more about them. Interactive whiteboards can be used to gradually reveal images and students can try to guess what the image depicts as it is gradually revealed.

This can lead on to discussion about how science works and the application and implications of science by consideration of the relationship between technology and science and the way that developments in microscope design and microscopy techniques have been so important in helping biologists to understand the cellular basis of life. Robert Hooke (1635–1703) was a scientist who developed scientific instruments and apparatus to extend human perception. He developed an improved microscope and in *Micrographia* (1665) was the first person to use the term 'cell' in a biological context. Hooke and his contemporaries believed that cells were only found in plants and it was not until the 1830s that cells began to be considered as fundamental units of life. The website of the Nobel Prize has some very useful information about the history of microscopy with useful links to the Nobel winners responsible for some of the advances in microscope technique. To emphasise the relevance of microscopy to their lives and how scientific ideas change over time, students could work in groups to find out more about the development of microscopes and link this to advances in medical treatment.

■ The functions of each part of a microscope

Having whetted students' appetites for studying small things, the microscopes can be got out! Younger students will usually use microscopes with mirrors beneath the stage for reflecting light from a bench lamp or the sky (not direct sunlight) towards the object being viewed. Eyepiece lenses usually magnify ×10 and there are often three objective lenses with magnifying powers of ×4, ×10 and ×40. When first using microscopes it is sensible to use only the low and medium power objective lenses. You should stress to students that microscopes are complicated and delicate pieces of equipment. It is a good idea to spend a whole lesson helping students become familiar with the particular type of microscope that they will be using. A useful exercise is to ask students to look for similarities and differences between their microscope and a diagram of one from a textbook. The functions of each part should be explained and a set of 'rules for using microscopes' formulated jointly with the students. Table 1.2 shows the sort of thing you should be aiming for.

Table 1.2 A set of rules for looking after microscopes

Microscopes are expensive and delicate instruments. To look after them properly we must:
- carry them carefully with one hand beneath and the other supporting the body of the microscope
- place them away from the edge of benches and sinks
- never touch the lenses with our fingers or try to take the microscope apart
- remove slides only when the tip of the objective lens is well away from the coverslip
- use only the fine adjustment control when the high power lens is in use and rack away from the object when focusing
- leave the microscope set on the lowest power objective lens and the body (or stage) racked fully down so that no strain is put on the cogs that move these parts
- pick up microscope slides and coverslips (very carefully) by their edges so as to keep them clean. Broken slides and coverslips should be put into the bin for broken glass
- always cover the microscope when it is not in use
- ask the teacher for help if we think there is something wrong with our microscope or the lenses need cleaning.

■ Using microscopes for the first time

Initial studies should be of specimens that can be seen with the unaided eye and do not require mounting under a coverslip. This will help to ensure that students achieve success straight away. Suitable things to view include small crystals such as salt or sugar, strands of students' own hair or fibres from their clothes, prepared slides of whole small animals such as fleas, or parts of animals such as insect mouthparts or wings. Before viewing these objects with a microscope, students could make observations with the unaided eye, hand lenses, mounted magnifiers and low-power binocular microscopes. The best way to use a hand lens is to hold the lens about 3 cm from one eye and bring the specimen into focus close to the lens. This technique enables students to concentrate their attention on the specimen being observed.

An important aim when using microscopes for the first time is to impress on students the usefulness of microscopes for extending the power of our sense of sight. To help develop this idea students could write descriptions of what the objects look like when viewed without being magnified and then when magnified using the hand lens and microscope.

Instructions produced for using the microscopes should relate specifically to the particular microscopes that are being used. Table 1.3 lists some general teaching points that could be used to help students get clear images.

Table 1.3 Teaching points when using microscopes

- With the lowest power objective lens in place, adjust the angle of the mirror (or the built-in lamp) to obtain a uniformly bright (but not too bright) field of view.
- Place the microscope slide on the stage with the object to be viewed directly beneath the lowest power objective.
- Clip the slide to the stage. Doing this will help to keep the slide firmly in position and means that when it is slid over the stage to centre the object in the field of view (see below) it will not move too quickly.
- Looking from the side, use the coarse focus control to rack the low-power objective downwards (or the stage upwards) until the lens is as close as it will get to the slide. On some microscopes there is a stopping mechanism that prevents the lenses from touching the slide.
- Looking through the eyepiece lens, slowly rack the lens away from the slide until the object is in focus. It is good practice to try to keep both eyes open when looking through a monocular microscope.
- Slowly move the slide across the stage until the object being viewed is in the centre of the field of view. The image will move the opposite way to the object – this takes students a little while to get used to.
- Adjust the illumination to gain a better image. With either too little or too much light little detail will be seen. This is particularly important with very thin specimens, which may not be seen at all if the illumination is too bright.
- Rotate the nosepiece to bring the middle power objective lens into position. Most microscopes used in schools are parfocal, which means that if the object is in focus when viewed under one objective lens it will remain in focus when a different lens is used. If the object does not appear in the field of view when a higher power objective lens is moved into position this is usually because the object was not accurately centred in the field of view with the lower power lens. Students have a tendency to immediately adjust the focus if the object is not visible. They should be told not to do this. Instead, they should carefully move the slide to search for the object. When it appears it should again be centred and if necessary the fine focus control used to sharpen the image and illumination adjusted to improve the image contrast.

■ Making temporary mounts

When viewing biological material with a microscope it is usual practice to mount the specimen in water (or a stain) and place a coverslip on top. Temporary mounts that are prepared properly have just enough fluid to fill the space between the slide and coverslip. If too much fluid is used the coverslip floats on top of the specimen and moves around. If too little fluid is used the air bubbles that are left interfere with the image and may be mistaken for the specimen. The technique for making a temporary mount is illustrated in Figure 1.1.

Rather than trying to demonstrate this procedure to the whole class it is a good idea to demonstrate it to a small group of students who then have to demonstrate it to another small group. Whilst the students are explaining you can listen and watch what they are doing to ensure that the correct procedure is being followed.

Figure 1.1 a Technique for lowering a coverslip on to a slide. The tip of a mounted needle is placed on to a slide next to the specimen. One edge of the coverslip is placed on the slide with the opposite edge supported by the mounted needle. Slowly moving the needle in the direction shown by the arrow will lower the coverslip on to the specimen without trapping air bubbles. b Tissue paper can be used to soak up any excess water, as shown below. A stain (e.g. iodine in potassium iodide solution) placed next to the coverslip can also be drawn under the coverslip using this technique

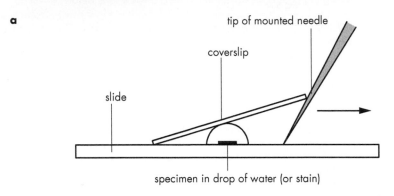

a

tip of mounted needle

coverslip

slide

specimen in drop of water (or stain)

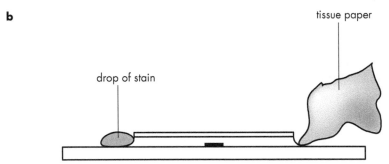

b

tissue paper

drop of stain

■ Helping students to see what you want them to see

When students begin to use microscopes they will need help in finding what you want them to see and in making appropriate observations. If they are not given help you will find that they mistake such things as dirt, air bubbles, the interesting patterns left when water or a stain evaporates and even the edges of the coverslip or slide for the things that you want them to look at. Drawing attention to the appearance of these artefacts is a useful way of helping students to avoid making this mistake.

Displaying images from a microscope using a data projector is an invaluable aid when teaching microscope techniques. There are a variety of ways in which images from light microscopes can also be displayed by a data projector. A small, traditional-style video camera can be fitted onto a microscope that has a vertical eyepiece 'tutor-tube'. Another technique is to use a 'flexicam' which has a camera on the end of a flexible 'gooseneck' arm. The camera can be inserted into an eyepiece tube and can also be used independently of a microscope. Digital eyepiece cameras are also available. A really useful piece of apparatus that can be used with an interactive whiteboard is a document camera or 'visualiser', which can also be used in conjunction with a microscope.

These systems can be used with students of all ages and abilities to demonstrate different techniques and to show images to a group of students. They can also be used to create a record of the images that students have looked at – this is very easy to do with an interactive whiteboard. These can be used as a starting point in the next lesson and a sequence of images can provide a valuable revision aid. To develop their communication skills students could combine such images with video footage gained using a hand-held digital video camera to produce short documentary-type films featuring their microscope studies.

Further activities

- It is a good idea to put clear plastic rulers on to the stage of a microscope and measure the width of the field of view at different magnifications using the millimetre divisions. The size of structures being viewed can then be estimated by judging how much of the field of view they occupy and doing some simple arithmetic.
- A huge variety of interesting microscopic organisms live in ponds or containers of water (e.g. cattle troughs) that have been left standing for some time. Study of these organisms can motivate students, develop their microscope techniques and demonstrate the usefulness of the microscope in biological studies. Students will be able to identify some of the microorganisms that they find but this is not of central importance. Students' communication skills could be enhanced by getting them to use these studies as a stimulus for some imaginative creative writing (e.g. poems, short stories, scripts for television or radio programmes) about the lifestyle of very small organisms.

 Good hygiene is needed.

Enhancement ideas

Further ideas for improving communication skills and helping students to appreciate the implications and applications of this aspect of science:

- Projects in which students have to use microscopic analysis of specimens such as hairs, fibres and powders to help solve a fictional crime can be very motivating. Students could work together as teams of investigators to produce an illustrated scientific report that could be used in the trial of a suspect.
- Microscopic studies provide many opportunities for making posters. These could include drawings and descriptions of microscopic organisms or could illustrate the development of light microscopes and their importance in many aspects of science.

1.3 Studying cells

Cell theory is an important unifying concept in biology. Briefly, it states that all living organisms are made up of cells and that all cells are derived from other cells. Many different organisms exist as single cells but most are made up of many cells. These multicellular organisms are composed of a variety of different types of cell that work together to maintain the life processes of the organism. You will find many references to cells in subsequent chapters of this book. This section suggests some teaching approaches that will help students to understand the structure of plant and animal cells.

Most cells are too small to be seen with the unaided eye so the use of microscopes is essential when studying cells. It is a good idea to wait until students are proficient at using microscopes before beginning practical work that is aimed at developing their understanding of the structure of cells. It is best to begin with plant cells because they are generally larger and have a more distinct structure than animal cells. Many sources of living plant material are easy to obtain and the techniques for handling them and mounting them on slides are relatively straightforward. Using living material will help students to relate the structures they look at under the microscope to the organism from which they came. In addition, preparing slides for themselves will help students develop their practical and manipulative skills.

Previous knowledge and experience

Some students will know that living things are made up of cells but they are unlikely to have studied this topic in any depth. Research has shown that younger students sometimes think the words 'molecule' and 'cell' have the same meaning and that this can give rise to a generalised concept of living things being made up of 'very small units' that can be molecules or cells. In addition, students sometimes think that many of the non-cellular things studied in the context of biology lessons (e.g. proteins, carbohydrates and water) are actually made of cells.

Some of the terms used in cell biology may be more familiar in other contexts, for example, cells in a honeycomb or a prison. The word 'cell' is also used in the context of a battery of electrical cells and when using spreadsheets on a computer. The difference between the nucleus of an atom and the nucleus in a cell will also need to be explained.

A teaching sequence

■ Studying plant cells

The easiest plant cells for students to study under the microscope are those that form single layers. The bulbs of onions and related species (e.g. shallots and leeks) are a good source of single layers of skin (or 'epidermal') cells. Red onions are particularly useful because the cells in the outer epidermal layers contain a red cell sap that makes them easier to see. It is relatively easy to interpret the details of each cell because they all lie in one plane and are not obscured by other overlapping cells. The technique of obtaining the epidermal cells, mounting them and viewing them under a microscope is described in student textbooks. It involves obtaining the epidermal layer by cutting an onion or leek in half and separating out the swollen leaf bases. The epidermal layer is peeled away using a pair of forceps, and a small pair of scissors is then used to cut off a piece measuring about 5×5 mm. If too large a piece of epidermis is used it is liable to become folded when placed on to a slide and, instead of being one cell thick, it will be two or more cells thick. To prevent tiny air bubbles being trapped next to the cells it is a good idea to wet the specimen by dipping it into some water before placing it on a microscope slide. Two or three drops of water or a stain are then put on the specimen and a coverslip placed on top using the technique illustrated in Figure 1.1 on page 10.

Plant cells with chloroplasts
One disadvantage of using onion epidermis as an example of plant cells is that these cells do not possess chloroplasts. The epidermal layer of privet, iris or lettuce leaves is a good source of cells that do possess chloroplasts. Whole leaves of Canadian pondweed (*Elodea canadensis*), ivy-leaved toadflax and young moss leaves are thin enough to enable cells to be seen when viewed under a light microscope.

■ Studying animal cells

Animal cells are not easy to see under the microscope so it is sensible to use images displayed on to a board to illustrate the structure of animal cells to younger students. Studying human cheek cells is an interesting and motivating activity to use with older students who are skilled at using microscopes. A cotton bud from a newly opened packet is used to collect the cells by gently swabbing the inside of the mouth around the gums. The saliva (which will contain the cells) is smeared on to a slide, a few drops of suitable stain (e.g. iodine in potassium iodide solution) are added and a coverslip is placed on top. After use, the cotton buds, slides and coverslip must be put into a disinfectant such as 1% sodium chlorate(I) (hypochlorite). The cells are very small and students will need to magnify them at least 100 times in order to see them clearly.

There is a very small risk of transmission of the viruses that cause AIDS and hepatitis B associated with this practical but most employers allow students to carry it out provided they follow a strict safety procedure.

Eye protection is needed when handling iodine solution.

■ Drawing and interpreting microscope images of cells

The drawing of images viewed using microscopes is an important skill in biology since it encourages careful observation and thus helps students to understand the images they are looking at. When drawing from the microscope, students should be taught to follow these procedures:

- Write a clear heading.
- Use a sharp HB pencil.
- Draw firm, continuous lines and avoid using too much shading.
- Include the magnification or a scale line.
- Draw label lines with a ruler. These must touch the structure they are labelling and should not cross each other.
- Write labels horizontally on the page (students often write along the same angle as the label lines) and arrange them neatly around (not over) the drawing.

Examples of good and not so good drawings are given in Figure 1.2.

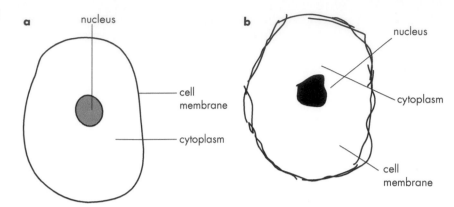

Figure 1.2 a Good and b Bad drawings of a human cheek cell

When they attempt to draw cells for the first time students often tend to draw too many cells. Instead they should be encouraged to draw a few cells in detail. To help interpret images of cells it is a good idea for students to draw what they can see, starting with the low-power objective lens, then use the middle- and high-power lenses. When using high-power lenses, students may be able to investigate the three-dimensional nature of larger cells (e.g. plant epidermal cells) by carefully focusing on different planes of the cell. Drawing or projecting diagrams on to a board and using models (see pages 18 and 28) will help students to interpret the structures of the cells they are viewing and see that these structures are three-dimensional.

■ Cell structure and function

You will want students to learn the names and functions of the main parts of cells. These are described in student texts. Some of the technical terms are difficult to spell and pronounce so it is worth going over them carefully.

Further activities

- Making three-dimensional models of cells using easily obtainable items such as shoe boxes, polythene bags and small balls to represent structures such as cell walls, membranes and nuclei is a very useful way of illustrating the structure of simple cells. Different types of model can be made using Plasticine, to illustrate the way in which cells become specialised (differentiated) for different functions. Students could also be asked to make cakes in the shape of different cells – edible homework!

- Since students often have to share microscopes, it is useful to adopt teaching techniques that take account of this, such as taking turns to use the microscopes. An activity that can be used to develop observational and descriptive skills that can be done in pairs or threes involves one student looking down a microscope and describing the appearance of a specialised cell (permanent cell mounts are best for this). The other student(s) have to draw and label what is being described. Encourage the students who are undertaking the drawings to ask questions to gain further information to enhance their drawings. They can try to predict what sort of cell is being looked at and then compare the drawings with the real thing. Students then swap roles and look at a different type of cell.

- Draw diagrams of plant and animal cells deliberately labelled incorrectly. Get the students to individually spot and make a note of the mistakes. Students then pair up and compile a joint list of errors. Follow up with a class plenary to correct all mistakes.

- Get all the students to draw a blank nine-square bingo grid. Write 12 key terms related to cell structure on the board and get the students to write nine of them into their bingo grid in any order they like. Read out definitions of the terms in a random order, e.g. by drawing out definition cards from a bag. Students cross off the terms when they match the definition. When a student calls 'line' they read out the terms and their meanings. Proceed to a 'full house' and again go over the definitions.

- To help students understand the concept of 'specialised cells' and the important idea that structure is related to function they could be given an assortment of easily obtainable items (e.g. staplers, scissors, forceps, cutlery) and asked to explain how their designs make them fit for purpose. Students could then be given lists of the different roles that cells perform and asked to draw a diagram of what a cell specialised to fulfil each role might look like.

- Give half the students in the class a card with a picture and description of a specialised cell on it. The other students have to ask questions which have a 'yes' or 'no' answer to work out what sort of cell the student has got on their card.

Enhancement ideas

- Pictures taken with scanning electron microscopes illustrate the external appearance, shape and three-dimensional nature of cells very clearly. Show students scanning electron microscope pictures of a selection of cells, e.g. red blood cells, unicellular organisms and the xylem cells that form wood, to demonstrate the variation shown by cells. Showing more able students the sort of images that can be gained using transmission electron microscopes (these show details of structures found inside cells) will help them to understand that cells are more complex structures than they appear to be when looked at using light microscopes.

- Simple micrometry ideas can be used to indicate to students the actual size of the cells.

1.4 The chemistry of life

In the first section of this chapter, life was described as arising from interactions between the chemical substances that make up organisms. Chemical reactions inside cells occur in sequences called metabolic pathways. These pathways involve either making or breaking down the large molecules that are found in living things. The steps along the pathway are catalysed by biological catalysts (enzymes) and the structure of each enzyme enables it to catalyse a particular chemical change.

Enzymes are proteins, so the temperature and pH of their surroundings affect their structure. If these vary outside certain limits then the shapes of enzymes change so much (i.e. the enzyme becomes 'denatured') that they are unable to perform their functions and metabolic pathways are disrupted. This is one reason why it is important for organisms to maintain relatively constant conditions inside their bodies and cells (i.e. show homeostasis). As well as producing products that are needed by organisms, metabolic reactions produce by-products that need to be removed. The removal of the waste products of metabolism is called excretion.

Previous knowledge and experience

The activities suggested below assume that students will have some knowledge of elements, molecules, compounds and the conservation of matter. Research has shown that students often think that molecules, which they have been told are 'large' (e.g. proteins), are bigger than cells, which they have been taught are very 'small'. Careful use of these relative terms is therefore needed. The names of some of the types of molecule found in living things will be familiar from adverts associated with food and cosmetics, while enzymes will often be associated with washing powders. Such advertising often contains somewhat questionable 'pseudoscience' that can give rise to many misconceptions, and students could be asked to use their scientific understanding to criticise the claims advertisers make for various products.

A teaching sequence

Examination specifications and student texts often include many details concerning the complex molecules that are needed by all living things, and enzyme action in the context of the human diet and digestion. However, before covering diet and digestion in detail, it is a good idea to go over some more general ideas about the chemistry of living things. The activities suggested below could be used in a variety of contexts and are not set out as a teaching sequence. It is hoped that the ideas suggested will enable you to help your students make links between many of the other topics discussed in this book and gain an overview of some of the overarching concepts of biological science.

■ Using models to develop ideas about metabolic pathways and the building up and breaking down of molecules inside cells

The idea that the complex molecules found in cells are assembled from smaller molecules is important in many contexts. For example, carbohydrates such as starch and cellulose are built up from simple sugars such as glucose; proteins are built up from amino acids; and nucleic acids are made up of sugars, phosphates and bases. The steps involved in the formation or breakdown of molecules in cells can be illustrated using models. This could be done using either physical models or a molecular-modelling package for use with computers. Models are also useful when relating the structure of a particular type of molecule to the functions that they perform in cells.

■ The role of enzymes

Student texts usually include a variety of investigations involving enzymes. Many of these are set in the context of digestion (see Chapter 2). Digestive enzymes are not typical enzymes since most enzymes are involved in reactions that take place inside cells, whereas digestive enzymes have their effect outside cells in the alimentary canal. Studying enzymes in the context of digestion can also give rise to the misconception that enzymes are primarily involved in the breakdown of large molecules into smaller ones, whereas far more enzymes are involved in catalysing the reactions that build up molecules. For these reasons it is important for students to carry out practical work that illustrates the role of enzymes in other contexts. For example, a useful demonstration is to show that an enzyme found in potato will catalyse the formation of starch from glucose. The enzyme is obtained by grinding up a small piece of potato (but not sprouting because it would have less enzyme) with water. The mixture is then centrifuged and a drop of the supernatant tested with iodine in potassium iodide (commonly called 'iodine solution') to ensure that it does not contain any starch. When the enzyme extract is mixed with glucose monophosphate the iodine test can be used again to show that starch is produced.

 Eye protection is needed when handling iodine solution.

■ Homeostasis and enzyme function

The importance of living organisms being able to maintain relatively constant conditions inside their cells (see Chapter 5) can be illustrated by carrying out practical work that illustrates how the rates of enzyme-controlled reactions are affected by pH and temperature. A variety of suitable investigations can be found in student texts.

■ Metabolic pathways and excretion

Excretion is often defined as being the release of waste products from living things. Two important ideas to emphasise when discussing excretion are, firstly, that the waste products released are made inside body cells and, secondly, that they will disrupt metabolism (i.e. are poisonous) if they accumulate inside the cells. A very common misconception is that undigested food material (faeces) is one of the excretory products of animals. Faecal material is not an excretory product because it is not made inside body cells and the correct term for the removal of faeces is 'defaecation' or 'egestion' although there are of course various alternatives!

Some metabolic waste products are molecules that can be used by the organism that produces them. For example, in plants the carbon dioxide produced by respiration can be utilised in photosynthesis and only at night is there a net release of carbon dioxide. Similarly, some of the oxygen produced by photosynthesis is used by plants for respiration. However, when the rate of photosynthesis is high, excess oxygen must be released. In contrast to plants, animals need to get rid of excess carbon dioxide all the time.

Animals also usually ingest more nitrogen than they need and so the other main metabolic waste products produced by animals are nitrogenous compounds, such as ammonia, that are formed by the breakdown of amino acids. Ammonia is very poisonous and in some animals is converted to urea, which is less toxic than ammonia, and filtered out of the bloodstream by the kidneys.

1.5 Energy and life

The energy needed for all living processes is made available inside cells by a metabolic pathway called respiration. This process takes place inside all living cells. Many students have difficulty understanding what is meant by the term 'respiration' and research has shown that they often retain many misconceptions after it has been taught, e.g. that 'respiration' and 'breathing' are synonymous, that plants do not respire or only respire when in darkness, and that respiration 'creates' energy for living processes.

There are two forms of respiration. Aerobic respiration occurs when sufficient oxygen is available and can be summarised by the equation:

food + oxygen → carbon dioxide + water + energy released

If there is insufficient oxygen available, anaerobic respiration occurs. For example, during strenuous exercise the blood system cannot supply oxygen to the muscles fast enough for aerobic respiration to be maintained. When this happens the muscle cells begin to respire anaerobically. This releases less energy than aerobic respiration and, in animals and some bacteria, can be summarised as:

food → lactic acid (strictly, lactate) + energy released

The lactic acid produced in muscles is broken down into carbon dioxide and water when enough oxygen is available to repay the 'oxygen debt' that results from anaerobic respiration.

In fungi (e.g. yeast), anaerobic respiration produces ethanol and carbon dioxide and can be summarised as:

food → ethanol + carbon dioxide + energy released

This is the fermentation process that is used in baking and brewing.

Previous knowledge and experience

Prior to being taught about respiration, students are most likely to have studied energy in the context of physics. They may have developed some understanding of the idea that energy is always conserved, but a very common misconception in a biological context is that energy is a physical substance that can be 'used up' (e.g. when we 'run out of energy').

The word 'respiration' itself is most likely to have been met in the phrase 'artificial respiration'. Artificial respiration should really be called 'artificial ventilation' because it actually involves the 'artificial' ventilation of a person's lungs. There is nothing 'artificial' about the respiration that one hopes will then ensue.

One of the difficulties in teaching about respiration is that although living organisms are respiring all the time, they do not show any obvious physical signs that this is occurring. Something that some animals obviously do is 'breathe', which is perhaps why respiration and breathing are often thought of as being the same thing.

A teaching sequence

The metabolic processes involved in respiration are invisible reactions that occur inside cells and cannot be studied at first hand in the school situation. However, many practical investigations concerning the reactants and products of respiration can be used to help students understand the process.

The fundamental importance of respiration to all living things can be emphasised by using a variety of different types of organisms in such investigations. Some possibilities are included in the teaching sequence set out below. Further ideas and more detailed explanations of the procedures that are suggested can be found in many student texts and on websites recommended at the end of this chapter.

■ Energy flow in living and non-living systems

Before teaching respiration, it is important to review students' knowledge of what is meant by the word 'energy'. We cannot see or feel energy – the physical experiences that we associate with it occur only when an energy transfer takes place. An 'energy circus' is a useful way of helping students appreciate that the concept of 'energy' can be applied in many situations. Students can be asked the question, 'Where is the energy now?' when presented with examples of living and non-living systems in which energy transformations are occurring. Examples of the ways in which living things utilise energy could include the production of movement, growth, sound, heat, light (in luminescent organisms) and electricity (as in nerve impulses or, more spectacularly, by electric eels). After considering situations in which a few energy transfers occur, students could then trace the many energy transfers that occur from the Sun, to their food, to their own life processes.

■ Food as a source of energy

Many foods and drinks are advertised as being 'high in energy', and studying a variety of food labels is a useful way of illustrating the point that food is needed to provide us with energy. Making the link between the high glucose content of 'high energy' foods and the photosynthesis that produces the glucose will help students to appreciate the idea of energy flow.

The idea that respiration involves food being 'burnt' inside the body to release energy is often used when introducing respiration. When doing this it should be made clear that respiration involves a series of small steps that transfer energy slowly, whereas burning is a much more rapid reaction. Dramatic demonstrations of the energy released when cornflour is burnt provide good stimulus material for introducing the idea that food provides the source of energy for animals (see websites listed at the end of this chapter for details of these demonstrations). Chapter 2 includes some ideas for measuring the energy content of different types of food.

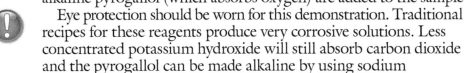

■ Demonstrating that oxygen is used in aerobic respiration

A simple demonstration to show that oxygen is used in respiration can be done by measuring how long a candle will burn in a gas jar of atmospheric air as compared with expired air. The composition of atmospheric and expired air can be demonstrated more accurately using a gas syringe to measure the reduction in volume that occurs when potassium hydroxide (which absorbs carbon dioxide) and then alkaline pyrogallol (which absorbs oxygen) are added to the sample.

Eye protection should be worn for this demonstration. Traditional recipes for these reagents produce very corrosive solutions. Less concentrated potassium hydroxide will still absorb carbon dioxide and the pyrogallol can be made alkaline by using sodium hydrogencarbonate, which is safer.

■ Demonstrating that carbon dioxide is produced in aerobic respiration

The well-known lime water test for carbon dioxide can be used when the amount of carbon dioxide being produced is relatively large, e.g. in air expired from the lungs. Carbon dioxide production by respiring plants and small animals occurs too slowly for the lime water test to give positive results. Instead, hydrogencarbonate indicator can be used to demonstrate that these organisms do produce carbon dioxide. Hydrogencarbonate indicator is a very sensitive pH indicator (details for how to make and use this very useful indicator can be found on the CLEAPSS website, the address of which is given at the end of the chapter). It has an orange/red colour when in equilibrium with the carbon dioxide levels of atmospheric air. When the carbon dioxide level rises slightly it changes to a yellow colour. This happens because the extra carbon dioxide that becomes dissolved in the indicator solution forms a very weak acid called carbonic acid. Placing pondweed (e.g. *Elodea*) into a boiling tube containing the indicator and excluding light to prevent photosynthesis will produce a colour change in about 1 hour. Pond snails are not harmed by the indicator and will cause a significant colour change in about 30 minutes. Leaves of terrestrial plants (e.g. privet) or small invertebrates (e.g. woodlice) can also be suspended above the indicator solution in the dark, producing a colour change after a few hours.

Data-logging equipment connected to pH probes is a very good alternative method of illustrating the pH changes associated with carbon dioxide production, especially if the respiration rate is very slow. Alternatively, use a carbon dioxide sensor.

■ Measuring heat production in germinating seeds

Germinating seeds convert energy from their food stores into the energy that they need for the metabolic reactions giving rise to growth. As in most energy conversion processes, heat is also produced. The temperature rise that occurs can be measured by placing a thermometer or a temperature probe (attached to a digital meter and if possible a data logger) into a thermos flask containing germinating seeds.

■ Anaerobic respiration in bacteria

Students are likely to be familiar with the smell or taste of milk that has 'gone off'. The sour taste results from bacteria in the milk respiring anaerobically and producing lactic acid. The fall in pH can be monitored over a number of days using a pH probe connected to a meter with a digital readout, which can be connected to a data logger. Comparing results obtained using pasteurised and long-life milk will illustrate the role of bacteria in this process.

■ Anaerobic respiration in yeast

Chapter 10 discusses some of the many practical possibilities for investigating anaerobic respiration (fermentation) in yeast.

■ Anaerobic respiration in muscles

The rapid exercising of muscles quickly results in a build up of lactic acid, and most students will be familiar with the sensations of tiredness, cramp and stiffness that follow strenuous exercise. A simple procedure that illustrates one of the reasons why these sensations occur is to clench and unclench your fists two to three times each second with one hand held above your head and the other by your side. The raised arm tires more quickly because it is harder to pump blood carrying oxygen upwards to the raised hand.

Further activities

- **Using respirometers to monitor oxygen uptake in small organisms**

  Respirometers can be used to measure oxygen uptake by small invertebrates or germinating seeds. The invertebrates or seeds are placed inside a container (e.g. a boiling tube) with a substance (usually pellets of soda lime) which will absorb the carbon dioxide produced by respiration. A bung connected to a U-shaped capillary tube with water inside is used to seal the boiling tube. The air pressure in the sealed container drops since oxygen is being used up in respiration and the water in the attached capillary tube is drawn along the tube. In practice this type of respirometer is fiddly to set up and may not provide reliable results unless it is completely airtight and maintained at a constant temperature. Depending on the ability of your students and the time you have available, it may be more sensible to use a respirometer as a demonstration rather than for a whole class practical activity. See also details about gas sensors on pages 55–56.

 Soda lime is corrosive.

- **A quantitative investigation of muscle fatigue**

  The following simple procedure provides a straightforward way of collecting quantitative data about fatigue induced by exercise. Students can then analyse these data to draw conclusions relating to the time it takes for an oxygen debt to be repaid.

PROCEDURE

1 A 200 g mass is suspended from the subject's index finger with the hand placed horizontally with the palm up over the side of a bench.
2 The mass is raised and lowered by bending the finger up and down as far as possible and as often as possible for 3 minutes. The number of times the mass is raised in each consecutive minute is recorded.
3 After resting for 5 minutes the procedure is repeated with 30 seconds rest after each minute of activity.
4 After a further 5 minutes rest the procedure is repeated with a 1 minute rest after each minute of activity.

Analysing the results should show that following periods of rest the finger muscles recover from the fatigue induced by the activity, and that longer rests result in better recovery. The recovery occurs because the 'oxygen debt' built up during the exercise (i.e. the amount of oxygen needed to get rid of the lactate) is repaid during the break from exercise.

- The appreciation that respiration is a fundamental process of life can be greatly enhanced by making explicit links between the study of respiration and many other topics. These could include movement that involves muscle contraction, movement of chromosomes during cell division, the synthesis of large molecules from smaller ones and the active transport of materials in and out of cells. Respiration is often studied in the context of gas exchange and ventilation, as discussed in Chapter 4. If this is the only context in which the process of respiration is discussed, the misconception that it does not occur in plants may not be addressed. For this reason it is particularly important to include a discussion of respiration when considering photosynthesis (see Chapter 2). The importance of respiration in the carbon cycle and in energy flow through ecosystems (see Chapter 9) should also be highlighted.

Enhancement ideas

- **The use of radioactive tracers in studying biochemical reactions**
  Much of our knowledge about metabolic processes such as respiration comes from experiments using radioactive tracers, e.g. following the path of a radioactively labelled carbon atom from glucose dissolved in an animal's drinking water to the carbon dioxide that the animal produces by respiring.
  With more able students some of the techniques involved and the results obtained can be discussed to help them understand the importance of such experiments in modern biology. Molecular models or 'student modelling', in which the students themselves represent different atoms (labelled and unlabelled), can be used to provide a physical representation of the process.

1.6 Cell division

Cell division is the basis of growth in multicellular organisms and of reproduction in all organisms. Students can be taught many things about growth and reproduction without discussing cell division in detail. However, to gain a deeper understanding of these processes, students will need to learn more about what happens when cells divide. The type of cell division that enables multicellular organisms to grow and to repair damaged tissues is called mitotic cell division or mitosis. Mitosis is also the basis of asexual reproduction, whereas sexual reproduction involves a different type of cell division called meiosis.

Teaching cell division presents a number of problems. It involves studying structures (chromosomes) which are very difficult to see even with the aid of high-powered light microscopes. The sequences of events that occur are complex and, when describing them, student texts often use many technical terms. More able students may be able to understand and use these terms but, before introducing them, it is a good idea to ensure that they understand the basic principles involved. The approach suggested here involves the use of video footage, diagrams and models to help students understand these principles.

Previous knowledge and experience

Before studying cell division students need to have some ideas about the relationships between chromosomes, genes and the characteristics that they determine. Some suggested approaches to this topic are given in Chapter 7.

A teaching sequence

■ Mitotic cell division

Mitosis should be covered before introducing meiosis. It is a fascinating process, and a good starting point is to show students a video of actual cells dividing and an animated sequence of this process (see websites listed at the end of the chapter for good sources of these). This can be followed by using a sequence of diagrams similar to that shown in Figure 1.3 to explain the important features of the process.

In order to explain the essentials of mitosis more clearly, this figure is a deliberately simplified representation of what really occurs. In all the diagrams the chromosomes are drawn as continuous or dotted lines with the circle around them representing the cell membrane (membrane and cell wall in the case of plants). In reality, when a cell is not dividing its chromosomes exist as very long, thin structures that are only just visible using electron microscopy. In the vast majority of species the chromosomes are contained within a nucleus. Stage 2 in the figure (when the chromosomes make copies of themselves) actually occurs when the chromosomes are inside the nucleus in their long, thin conformation. The nuclear envelope breaks down and the chromosomes become shorter and fatter (and visible under a light microscope) *after* this has taken place. Many other details have been omitted and few technical terms are used. You can find more detailed accounts of the process in advanced student texts. Some of the technical terms that you might want to use are included in the figure captions and are printed in italic script. You will need to judge when it is best to introduce them. For some students they are probably best avoided completely.

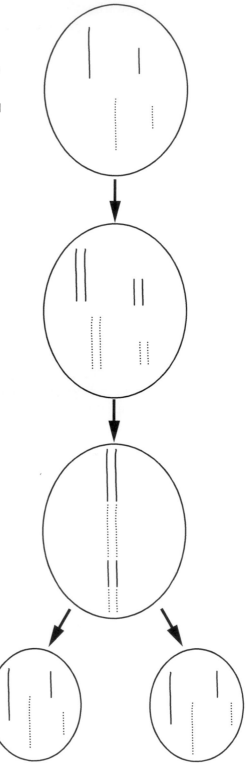

Figure 1.3 A diagram summarising the events that occur during mitosis

1 A cell containing two pairs of (*homologous*) chromosomes. One of each pair comes from the original female parent and is shown as a continuous line. The other comes from the original male parent and is shown as a dotted line.

2 Each chromosome makes an exact copy of itself. (The copies are called *chromatids*.)

3 Chromosomes (now pairs of *chromatids*) line up in the middle (on the *equator*) of the cell.

4 The copies (*chromatids*) separate and move to opposite ends (*poles*) of the cell, which then divides to form two (*daughter*) cells. Each of these has exactly the same genetic make up as the original cell.

Rather than using pre-prepared diagrams it is a good idea to draw diagrams similar to those shown in Figure 1.3 on to a board or overhead transparency during a lesson using different colours (rather than continuous and dotted lines) to represent the chromosomes. Writing on to an overhead allows more eye contact with students, which makes it easier to judge an appropriate pace for your explanation. It is essential that students concentrate on the explanation rather than try to copy diagrams or write things down whilst you are talking.

Once the sequence has been completed, students' understanding can be reviewed by asking them questions about each of the stages you have illustrated. A crucial point to establish is that the cells formed as a result of mitotic division contain exactly the same complement of chromosomes (and hence genes) as the original undivided cell.

A disadvantage of using diagrams to illustrate mitosis is that it is difficult to show the behaviour of the chromosomes in three dimensions. Commercially produced models can be used to help students visualise the three-dimensional nature of the process. Students could also use their own plasticine or wool models to aid their understanding. Following these activities, the act of drawing diagrams similar to those in Figure 1.3 will help students to consolidate their understanding. A useful idea is to provide them with written descriptions of each stage that they then illustrate diagrammatically. Finally, you could return to the video sequence with which you began and review the whole process.

■ Meiosis and gamete formation

The type of cell division that produces sex cells (gametes) is called meiosis. Gametes differ from other types of cell (somatic cells) produced by organisms in two fundamental respects. First, they contain half the number (the haploid number or 'n') of chromosomes that are present in the somatic cells (the diploid number or '$2n$'). If this were not the case the zygotes (fertilised eggs) formed when sex cells combine (fuse) would contain twice as many chromosomes as the parents' cells.

Secondly, many of the chromosomes present in gametes contain a mixture of genes derived from the male and female parents of the individual that is producing the gametes. Figure 1.4 illustrates the important events that occur during meiosis. In common with Figure 1.3, this is a deliberately simplified illustration of what actually happens. Some of the technical terms you might wish to use with students are printed in the captions in italic script. Comparison with Figure 1.3 shows that the initial stages of meiosis and mitosis are essentially the same. A crucial difference occurs

Figure 1.4 A
diagram
summarising the
events that occur
during meiosis

1 A cell containing two pairs of
(*homologous*) chromosomes.
One of each pair comes from the
original female parent and is
shown as a continuous line.
The other comes from the original
male parent and is shown as a
dotted line.

2 Each chromosome makes an
exact copy of itself. (The copies
are called *chromatids*.)

3 Equivalent (*homologous*) pairs of
chromosomes (each made up of two
copies or *chromatids*) line up
alongside each other in the middle
(on the *equator*) of the cell.

4 Equivalent (*homologous*)
chromosomes move to
opposite ends (*poles*) of the
cell and two new cells
are formed.

5 The chromosome copies
(*chromatids*) now separate
and move to opposite
ends (*poles*) of each cell.
These divide forming a
total of four new cells.
The number of
chromosomes in each
is half the number found
in the original cell.

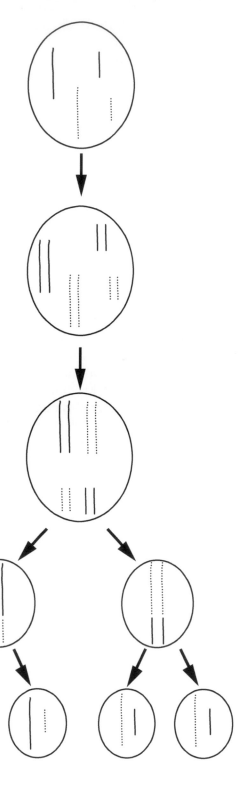

after the chromosomes have made copies of themselves. In mitosis, the chromosomes become arranged in a single line in the middle of the cell, and then separate. In meiosis, the chromosomes line up in pairs with equivalent (homologous) chromosomes from the male and female parent of the organism becoming aligned next to each other in the middle of the cell. These then move to opposite ends of the cell, which then divides. Another division follows in which the chromosome copies (chromatids) separate and a total of four cells is formed. The cells that result from meiosis therefore contain half the number of chromosomes found in the original cell.

The gametes that these cells produce will be genetically different from each other because, at the stage in meiosis when the chromosomes pair up, some will have chromosomes derived from the male parent on one side of the pairing, whereas others will be the opposite way round. This is referred to as 'independent assortment' of the chromosomes and leads to independent segregation of alleles (see Chapter 7). Figure 1.4 shows one of the two possibilities that exist when there are two pairs of chromosomes in the original cell. The other possibility is for both of the chromosomes derived from the maternal parent (represented by continuous lines) to be on the same side of the pairings.

Additional variation between gametes occurs because genetic material is exchanged between chromosomes when they pair up with each other. This, together with the variation resulting from independent segregation of chromosomes, results in an almost infinite amount of genetic variation between the gametes produced by an individual (see Chapters 6 and 7). This contributes to the genetic variation that occurs in the offspring produced by sexual reproduction and this variation has important consequences for the evolution of species, as discussed in Chapter 8.

Further activities

- An activity in which students play the role of chromosomes and enact cell division provides a useful way of reinforcing learning about this process. To provide guidance on how you might do this go to 'YouTube' where you will find various approaches, including one involving synchronised swimming!
- Students could use computer drawing packages to produce diagrams or PowerPoint presentations illustrating mitosis. The simple set of instructions on the facing page could be used as guidance:

1 Draw a diagram of a cell containing one pair of chromosomes, using a different colour for each.
2 Chromosomes make exact copies of themselves.
3 Chromosomes become shorter and thicker.
4 Chromosomes line up in the middle of the cell.
5 Each copy moves to opposite ends of the cell.
6 Two new cells are formed.

The procedures that students will use to create their diagrams model some of the events that occur during actual cell division, e.g. using 'copy and paste' to make copies of chromosomes, changing their size and shape and moving them to different positions within a cell.

- Getting pairs of students to model the sequence of events using pipe cleaners of different colours to represent chromosomes is another useful technique that can be used to encourage discussion of the processes involved in cell division and thus reinforce understanding.

Enhancement ideas

- Older students with good microscopy skills could look at prepared microscope slides showing the different stages of mitosis and meiosis. They could also prepare root-tip squashes (see the SAPS website below for practical details) to look for different stages of mitotic division.

Other resources

Generic websites
The website put together by Don Mackean (author of many biology textbooks) contains many helpful resources including PowerPoint presentations on topics such as 'Cells and tissues', 'Cell division and specialisation', 'Chromosomes' and 'Respiration'. There are also suggestions for practical activities linked with each of these topics, diagrams and question banks. Student worksheets and teachers' guides related to many of the activities suggested in this chapter can also be found at this site: http://www.biology-resources.com/.

eChalk provides excellent diagrams, animations and images for use with an interactive whiteboard. Subscription is needed to access these resources: http://www.echalk.co.uk/Science/biology.aspx.

'Practical Biology' is a website that is a joint project of the Nuffield Curriculum Centre, the Society of Biology and CLEAPSS, and is intended to encourage teachers to carry out more practical biology in schools: http://practicalbiology.org/.

The 'Eurovolvox' website provides a growing range of resources for biology teaching. Images on this site relating to life processes, levels of organisation, cells and the chemistry of life can be used to help create stimulating PowerPoint presentations: http://www.eurovolvox.org/menu.html.

Science and Plants for Schools (SAPS) has a range of materials and suggestions for practical work, including investigating mitosis in root tips and viewing microorganisms using the 'hanging drop' technique: http://www-saps.plantsci.cam.ac.uk.

'Teacher Resource Exchange' is a site that has many useful resources and ideas (including lesson plans, PowerPoint presentations and simulations) related to the topics in this chapter: http://tre.ngfl.gov.uk/.

'Kings Science' – Richard King has put together a very useful set of free resources, including flash animations and worksheets: http://www.kscience.co.uk/index.htm.

The Chelmsford County High School website provides high-quality matching exercises, labelling diagrams, ordering diagrams and many other useful resources: http://www.cchs.co.uk/subjects/biology-ks3online.php.

The 'Science Photo Library' site provides a vast number of high quality photographs – the images of cells and tissues are excellent: http://www.sciencephoto.com.

Practical instructions for a cornflour bomb demonstration to indicate the energy released in combustion/respiration can be found at http://www.practicalchemistry.org/experiments/the-cornflour-bomb,224,EX.html and a demonstration of how to burn cornflour with dramatic results at http://www.youtube.com/watch?v=evXhs1-exMo.

Websites related to microscopy

The website of the Royal Microscopical Society has an 'outreach' section that includes advice on buying microscopes for use in schools. It also has details of a 'Microscope Loan Box Scheme' through which schools can loan eight microscopes, a digital camera, activity workshops and samples: http://www.rms.org.uk/.

'BSI Education' is the website of the British Standards Institution (BSI). It has very useful material on the history of microscopy, methods for setting up and adjusting a light microscope and for preparing temporary mounts. It also has suggestions for assignments using microscopes that students could carry out: http://www.bsieducation.org/Education/14-19/topic-areas/applied-science/microscopy/default.shtml.

The website of the Nobel Prize has some useful material on the development of microscopes, the use of different types of microscopes and a game related to the cell cycle: http://nobelprize.org/educational/.

The website 'Microscopy UK' includes some very good images and videos suitable for use in PowerPoint presentations or as stimuli for discussions, together with resources designed specifically for use in schools: http://www.microscopy-uk.org.uk/.

Nikon has developed a very useful website devoted to microscopy education. Although much of the material available is aimed at degree-level and research scientists it is worth visiting the site to find amazing images (in the 'Small World' competition winners and the digital image galleries), details of the development of microscopy techniques (in the microscopy museum) and video sequences of living cells moving around (in the live cell imaging section): http://www.microscopyu.com/.

The CLEAPSS website has a very large range of material devoted to practical science, including detailed instructions for using microscopes: http://www.cleapss.org.uk.

'Bioscope' is a computer simulation of a real microscope with 56 plant and animal microscope slides at a range of magnifications, appropriate for use with students at various levels. It is produced by the University of Cambridge Local Examinations and details can be found at http://www.cambridge.org.

The 'Open University Online Digital Microscope' also has a variety of images of specimens with accompanying notes that can be viewed online. It can be found at http://www.open2.net/science/microscope/frames.html.

An astonishing demonstration of writing on a human hair was shown on the BBC programme 'Invisible Worlds' featuring Richard Hammond: http://www.bbc.co.uk/programmes/b00rt87x clips.

Websites related to cell division

Putting the search term 'cell division' into 'YouTube' will take you to many potentially useful videos, including modelling of the process of cell division by synchronised swimmers, songs about the process, as well as videos of the actual process. Some of the best videos of the process of cell division have a commentary that may be too detailed, so it is useful to provide your own explanation of the process that suits the age and needs of your students.

The following two sites have clear animations of mitosis and meiosis: http://www.johnkyrk.com/mitosis.html, http://www.johnkyrk.com/meiosis.html.

'Gene Journey' is a regularly updated online resource on genetics that includes animations to explain the processes of mitosis and meiosis. It also includes multiple choice quizzes and interactive games: http://www.illumination-ed.co.uk/.

Teaching aids

Video cameras, digital microscopes and USB microscopes are available from many suppliers and are ideal for displaying images from a microscope on to a screen using a data projector. Software for capturing, analysing and logging images that are obtained with these cameras can be used in a variety of exciting ways with students. Dino-lite digital microscopes produce a range of digital microscopes and associated software specifically designed for use in schools: http://www.dinolite-uk.com.

A variety of three-dimensional models illustrating cell structure, important biological molecules and the stages of mitosis and meiosis are available from school science suppliers.

Visits

Many museums have displays illustrating the development of optical microscopes and microscopy techniques. For schools in the London area a visit to the exhibition called 'Centre of the Cell' is strongly recommended. This won the Educational Initiative award at the 2010 Museums & Heritage Awards for Excellence. Details can be found at http://www.centreofthecell.org/, which also has useful interactive learning resources and an online shop.

Schools in the Oxford area could visit the 'Museum of the History of Science', which offers sessions designed to give students an insight into the impact of the invention of the microscope in the seventeenth century with an opportunity to look at historic books from the library, including an early edition of Robert Hooke's *Micrographia* (1665) and to see a range of microscopes illustrating the way in which technological developments have enabled biologists to find out more about cell structure and function: Details can be found at http://www.mhs.ox.ac.uk/

Further reading

Batts, G.R. (2000). Cell and school activities – a useful analogy, *School Science Review*, **81** (296), pp. 111–112. A paper describing an analogy that compares and contrasts organelle activities with activities necessary to maintain a functioning and efficient school.

Dreyfus, A. and Jungwirth, E. (1989). The student and the living cell: a taxonomy of dysfunctional ideas about an abstract idea, *Journal of Biological Education*, **23**, pp. 49–55.

Forrest, B. (2003). Designing an organism, *School Science Review*, **84** (309), pp. 11–12.

Kinchin, I.M. (2000). Star Wars biology: Why are Wookies so hairy? *School Science Review*, **81** (296), pp. 109–110. Suggestions for using science fiction and films to generate enthusiasm when introducing the concept of adaptation.

Lester, A. and Lock, R. (1998). Flexible visual aids – sponge enzyme models, *School Science Review*, **79** (289), pp. 105–107.

Lock, R. (1991). Creative work in biology – a pot-pourri of examples. Part 1 – Expressive and poetic writing, cartoons, comics and posters, *School Science Review*, **72** (260), pp. 30–46.

Lock, R. (1991). Creative work in biology – a pot-pourri of examples. Part 2 – Drawing, drama, games and models, *School Science Review*, **72** (261), pp. 57–64.

Marsh, G., Parkes, T. and Boulter, C. (2001). Children's understanding of scale – the use of microscopes, *School Science Review*, **82** (301), pp. 27–31.

Martin, M. (2002). Looking at cells: alternatives to onion cells, *School Science Review*, **83** (304), p. 103.

Tregidgo, D. and Ratcliffe, M. (2000). The use of modelling for improving pupils' learning about cells, *School Science Review*, **81** (296), pp. 53–59. A report of a study which compared the use of 2D and 3D modelling in improving pupils' learning about cell structure and function.

Lewis, J., Leach, J. and Wood-Robinson, C. (2000). Chromosomes: the missing link – young people's understanding of mitosis, meiosis, and fertilisation, *Journal of Biological Education,* **34** (4), pp. 189–199.

Lewis, J. and Wood-Robinson, C. (2000). Genes, chromosomes, cell division and inheritance – do students see any relationship? *International Journal of Science Education,* **22** (2), pp. 177–195.

Snedden, R. (2007). *Cells and life: The world of the cell*, Oxford: Heinemann.

Snedden, R. (2007). *Cells and life: Cell division and genetics*, Oxford: Heinemann.

These two last books are useful sources of information for non-specialists teaching biology as well as for students studying biology.

2 Nutrition, diet and photosynthesis

Neil Millar

2.1 Diet
- Food groups
- Energy in food
- Balanced diet
- Diet and health

2.2 Digestion
- Human digestive system
- Ingestion
- Digestion
- Digestive enzymes
- Absorption
- Egestion

2.3 Photosynthesis
- Plant growth
- Leaves
- Testing for starch
- Testing for oxygen
- Limiting factors
- Photosynthesis and respiration
- Testing for carbon dioxide

Choosing a route

This chapter is about animal and plant nutrition. Although linked by the idea of nutrition, one of the fundamental characteristics of life, the methods of nutrition in animals and plants are so different that the two topics are best taught separately.

We start with animal nutrition, which will be the more familiar to students, particularly as the focus is mostly on humans. We consider what humans eat, the concept of a balanced diet and the health problems associated with a diet that is unbalanced. We then move on to what happens to food once it is eaten: how and why it is digested and absorbed into the body. Plant nutrition is covered in the third section. Plants don't eat, but they do feed themselves. They make their own food from air and water in the process of photosynthesis, using energy from the Sun.

The topics can be taught in any order. Many opportunities for investigative work are described throughout the chapter. The object of these practical investigations might be to help teach the scientific knowledge or it might be to teach the skills associated with the scientific method, such as experimental design, data presentation and data interpretation. In addition, the practical work ought to enthuse and actively involve the students in the subject. The practical work can be carried out at any time of year, since the plants used can be obtained from garden centres and aquarium suppliers throughout the year.

2.1 Diet

Previous knowledge and experience

By age 11 students will be quite familiar with many aspects of diet both from everyday experience and from primary school. They will know about a balanced diet and about food groups such as meat, dairy, fruit and vegetables, and some will have heard of carbohydrates, proteins, fats and fibre. They may be aware of different weight-loss diets, diet advice (such as 'five a day') and diet 'gurus' from magazines and TV. This previous knowledge is helpful, but some incorrect ideas may have to be addressed as they arise.

A teaching sequence

A good starting point for work on food is to explore students' existing knowledge by asking, 'Why do we eat food?' The various suggestions can be discussed and can probably be summarised in just two reasons:

- to give us raw materials for growth and repair
- to give us energy.

All organisms need matter and energy from their environment, and animals obtain both from their food (recall that the arrows in a food chain represent the flow of matter and energy). The energy is obtained by 'burning' the food in the process of respiration. The two uses of food lead into the idea of food groups.

■ Food groups

Students' knowledge of food groups can be found by asking, 'What different foods do we eat?' Students will probably come up with suggestions such as fruit, vegetables, meat and dairy, which are useful groupings but not very scientific. The more scientific terms are the seven food groups: carbohydrates, proteins, fats, minerals, vitamins, fibre and water. Students could research good sources of these seven groups, and what they are needed for, producing leaflets or posters.

The UK food standards agency website has the 'eatwell plate' (Figure 2.1), which illustrates the food groups, and the British Nutrition Foundation is also a good source of dietary information.

Figure 2.1 The 'eatwell plate' (from the UK Food Standards Agency)

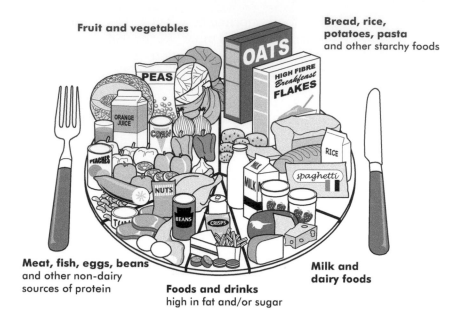

Fruit and vegetables

Bread, rice, potatoes, pasta and other starchy foods

Meat, fish, eggs, beans and other non-dairy sources of protein

Foods and drinks high in fat and/or sugar

Milk and dairy foods

■ Food tests

The concept of food groups can be reinforced by practical work here, as students test real foods themselves for the presence of some of the food groups. There are simple qualitative tests for starch (iodine solution turns blue–black), sugar (Benedict's solution turns red), protein (biuret solution turns lilac), fats (an ethanol solution of the food turns milky when decanted into water) and vitamin C (DICPIP solution is decolourised). The difference between reducing and non-reducing sugars is not needed at this stage. Details of the tests can be found in many textbooks and websites such as Mackean's Biology Resources. Students can test a variety of raw and processed foods, such as rice, milk, cereals, pasta, biscuits, onions, beans, apples, etc., and fill in a suitable table to show their results. Discussion of the results obtained can throw up some surprises (e.g. onions contain sugar but not starch).

■ Energy in food

To illustrate that food contains energy it can be burned, showing that food is a fuel just like petrol or wood. Some foods will burn easily on a tripod and gauze and can be used to heat water in a boiling tube above. Only a few high-fat foods will burn successfully in air this way, but at least students can do it themselves.

 Nuts should never be burned as students may have nut allergies.

Alternatively, food can be burned in a laboratory food calorimeter, which burns samples of food in an atmosphere of pure oxygen.

The food calorimeter will give better results, as almost anything will burn in it and it's more like the real bomb calorimeters that food manufacturers use to obtain the energy data printed on food labels (Figure 2.2).

Figure 2.2 Food calorimeter

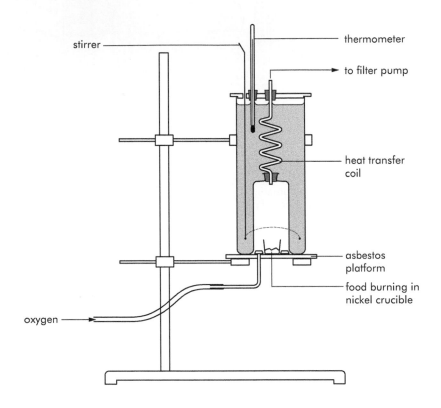

stirrer

thermometer

to filter pump

heat transfer coil

asbestos platform

food burning in nickel crucible

oxygen

The chemical energy stored in the food is given off as heat, heating the water. The hotter the water gets, the more energy in the food. For more advanced work, students can use the following formula to convert the temperature rise to energy content and then compare the results to the food label.

$$\text{energy (kJ per 100 g)} = \frac{\text{temperature rise (°C)} \times \text{water volume (cm}^3) \times 4.2}{\text{mass of food (g)} \times 10}$$

The food calorimeter makes an excellent exercise in evaluation. Was it a fair test? Only if we controlled factors such as volume of water, initial temperature of water, mass of food, stirring, rate of oxygen flow, etc. Was it accurate? The main errors were heat loss through the vented gases and through the water. How could the accuracy be improved? By using a longer coil and better insulation.

■ Analysing the diet

Students can analyse their own diets in terms of food groups and of energy using software such as the Diet Analyser unit in Multimedia Science School. Students first record a 'food diary' of everything they ate over a 24-hour period, including estimated quantities. Once they have entered this into the analyser software they get a table and chart showing how much of each food group they ate, compared to their recommended daily allowance (RDA). Although this activity is excellent for bringing home the idea of the different food groups and relating them to personal experience, a great deal of sensitivity is needed. Some students may be concerned that their diet appears unhealthy, and you must stress that this is just a rough-and-ready exercise, for fun, based on only 1 day's eating and using educational software. The results of this activity could be wildly inaccurate, and a serious analysis would require accurate weighing and recording over a period of at least a week, with analysis by a trained dietician.

A more straightforward exercise is simply to analyse the composition of different foods by studying food labels and food data tables in books. These analyses will show that while many foods contain a variety of different food groups, certain foods are associated with specific food groups (e.g. bread and pasta with starch, meat with protein, fruits with vitamin C and processed food with fat and sugar).

The work so far can be summarised in the concept of the 'balanced diet'. This term can be confusing, since it refers to two different kinds of balance: there should be a balance of different food groups (such as illustrated in the eatwell plate shown in Figure 2.1) and there should also be a balance between the energy consumed and energy used. Both aspects of this balance mean that diets will be different for different people, and students should be able to suggest how diets will be affected by factors such as age, gender, physical activity, job, diet choice (such as vegetarians and vegans) and medical issues (such as food allergies).

■ Diet and health

With older students it may be appropriate to consider the effects of an unbalanced diet on health in more detail.

This can be quite a sensitive area, and you should be aware of any students with known eating disorders or medical issues in the class. There is no appropriate practical work, but this topic lends itself to students' own investigations. Students can research topics in books and on the internet and prepare leaflets, posters or presentations of their findings. Presenting information and explaining concepts to others are useful skills to develop, and some students can play the role of 'critical friend', evaluating their peers' presentations. The NHS website is a good place to start, and many health companies in the USA have excellent websites with articles, diagrams and animations. The choice of topics will depend on the syllabus being followed and on the age and ability of the students.

An unbalanced diet causes malnutrition, which can take many forms. If the energy content of the food taken in is less than the amount of energy used then the person loses mass and may become underweight. This can happen in the developing world when a balanced diet is not available, and in the developed world if someone is on a weight-loss diet. Conversely, if more energy is taken in than used then the person gains mass and may become overweight. This can happen in the developed world when the easy availability of high-calorie food is accompanied by a sedentary lifestyle. What is a normal weight? This depends on many factors, but the simplest measure is the Body Mass Index (BMI), which is given by the formula:

$$BMI = \frac{\text{mass (g)}}{\text{height (m)}^2}$$

An adult whose weight is normal for their height has a BMI in the range 18–25. Less than 18 means underweight and more than 25 means overweight. A BMI greater than 30 means obese, where excess body fat has accumulated so much that it is likely to have an adverse effect on health. For children, the normal BMI range depends on sex and age, and charts showing this can be found on Wikipedia. Students should not be asked to calculate their own BMIs in class for fear of stigmatising students whose BMIs fall outside the normal range. Instead they can calculate BMI values for given fictitious individuals and be asked to 'diagnose' any weight problems and suggest solutions.

Other diet and health topics might include:

- **Deficiency diseases:** these are more common in developing countries where food supplies may be erratic and diets lack variety. It is estimated that 30% of the world's population is at risk of developing iodine deficiency and at least three million children are blind due to vitamin A deficiency. Other deficiency diseases include scurvy (lack of vitamin C), rickets (lack of vitamin D or calcium), anaemia (lack of iron) and kwashiorkor (lack of protein).

- **Cholesterol:** is needed by the body to make cell membranes and steroid hormones, and, like all fats, is highly insoluble in water. High concentrations of cholesterol can therefore cause blockage of blood vessels and lead to high blood pressure and heart disease. Cholesterol is transported in the blood as lipoproteins. Low density lipoproteins (LDLs) carry cholesterol from the liver (where cholesterol is made) to other organs and increase the risk of heart disease, while high density lipoproteins (HDLs) carry cholesterol from other organs to the liver and decrease the risk of heart disease. LDLs are known as 'bad cholesterol' and should be kept low, while HDLs are known as 'good cholesterol' and should be kept high.
- **Saturated and unsaturated fats:** fats from warm-blooded animals (in meat and dairy foods) tend to have saturated fatty acids. These raise blood cholesterol so should be kept low. Plant and fish fats (olive oil, sunflower oil, oils from oily fish) tend to have unsaturated fatty acids. These lower blood cholesterol and LDLs, so should be chosen in preference to saturated fats. Polyunsaturated fats are better than monounsaturated fats. Omega-3 fatty acids are a particular group of unsaturated fatty acids found in fish oil that have putative health benefits.
- **Slimming pills:** students can evaluate data on the effectiveness of slimming pills, and investigate the causes of eating disorders such as anorexia nervosa and bulimia.
- **Food additives:** students can investigate the purpose and prevalence of E numbers, including food colourings. E numbers may also be covered in chemistry lessons.
- **Food emergencies:** it can be interesting to research diet problems and solutions associated with famines and other emergencies, such as the use of 'plumpy-nut'.

2.2 Digestion

Previous knowledge and experience

In contrast to diet, students will probably have far less knowledge of digestion. They may know the locations of some parts of the digestive system, and they probably have a vague notion that the 'goodness' is taken out of food and the rest removed as faeces. Few will have any idea of chemical digestion or enzymes.

A teaching sequence

The obvious sequence is to follow the fate of food that is eaten. There are four stages:

- ingestion – taking large pieces of food into the body
- digestion – breaking down the food by mechanical and chemical means
- absorption – taking up the soluble digestion products into the body's cells
- egestion – eliminating the undigested material.

First though, students will need to know the structure of the human digestive system. As so often in biology there is a good link between the structure of an organ and its function.

■ The digestive system

Students will need to learn the names and locations of the main parts of the human digestive system. Models, videos and software can all be used to help demonstrate the anatomy, and it may be appropriate to dissect a preserved animal, such as a rat, to show a real digestive system. Teachers TV has a video of a teacher in a rural school dissecting a pig's gut, which certainly makes an enthralling learning experience. Key points include:

- The digestive system comprises a long tube, called the alimentary canal, digestive tract or gut, which extends from the mouth to the anus.
- The gut is lined with muscles, which contract in waves to push the food through the gut. This peristalsis can be demonstrated by pushing a ball through a soft tube like a jumper sleeve.
- The stomach, although well known, is actually one of the less important parts. Food is stored and churned here, but not much digestion takes place. The stomach contains hydrochloric acid to kill bacteria that are living in the food. You can feel the acid burning your throat when you are sick.
- The most important processes (digestion and absorption) happen in the small intestine, which is very long (8 m in an adult). The length can be demonstrated with an 8-metre ribbon or piece of string.
- The digestive system also includes three glands: the salivary glands, the liver and the pancreas. Food does not pass through these, but they secrete digestive juices into the gut to aid digestion.

■ Teeth

Teeth are generally not studied as much as they used to be, but can be an interesting topic for students. Teeth are the main tools of mechanical digestion – physically breaking up large, tough chunks of food, which mix with saliva to form a slurry of tiny fragments. Humans and other mammals have four different types of teeth, each specialised for a different job. Incisors are used for cutting, canines are used for tearing and premolars and molars are used for grinding. Students can fill in a dental chart (which can be found in textbooks or on the internet) of their own teeth to see how many of each they have. Images and animations can help to show the structure of a tooth and how it can be damaged by decay. If available, model or real skulls of a sheep and a dog can be used to compare the teeth of herbivores and carnivores. The different structures of the jaws and teeth can be related to their different diets.

■ Digestion

An experiment with Visking tubing can be used to demonstrate the need to digest food. The Visking tubing is a 'model gut': inside is the food and outside is the bloodstream. A 15 cm length of Visking tubing is filled with a mixture of starch solution and glucose solution. The outside of the Visking tubing is rinsed and placed into a boiling tube of water for about 30 minutes at 40 °C (Figure 2.3). The contents of the Visking tubing and the water in the boiling tube are then tested for starch (using the iodine test) and for sugar (using the Benedict's test). The results will show that both starch and glucose are present inside the tube (as expected), but only sugar is present in the water outside the tube.

Figure 2.3 Using Visking tubing as a model gut

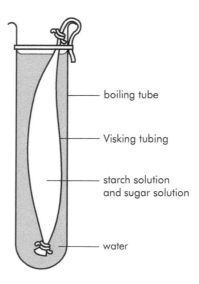

Students need to be told that Visking tubing is like a sieve with very small holes that will let small molecules through, but not large ones. Then they can make two conclusions:

1 Starch cannot get through Visking tubing, so its molecules must be too big.
2 Sugar can get through Visking tubing, so its molecules must be small.

This demonstrates the purpose of digestion: to break down large, insoluble molecules into small, soluble molecules that can pass through the wall of the gut into the bloodstream. Strings of beads or Lego can be used to demonstrate large insoluble molecules being broken down into small soluble molecules, so they can fit through the holes in a net bag (Figure 2.4).

Figure 2.4 A model of digestion and absorption

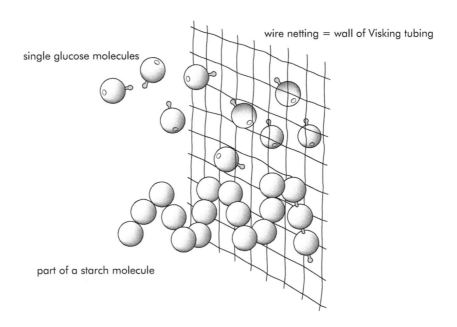

wire netting = wall of Visking tubing

single glucose molecules

part of a starch molecule

Students should realise that not all food is digested:

- Small molecules are soluble and don't need to be digested. They are already able to cross the gut wall into the blood, e.g. sugars, minerals, vitamins and water.
- Large molecules are insoluble and must be digested to small, soluble molecules before they can get into the blood, e.g. starch, proteins and fats.
- Fibre is a large molecule that we can't digest. It cannot cross the gut wall, so remains as faeces, e.g. cellulose plant cell walls.

■ Digestive enzymes

Large molecules are digested to small ones using digestive enzymes. This may be students' first encounter with enzymes. They may have come across catalysts in chemistry lessons and so should understand that enzymes are biological catalysts. Students generally need to know about three main digestive enzymes, including what they do and where they are made:

- The enzyme **amylase** digests starch molecules into sugars. (Older students may be taught that starch is first digested to maltose and then to glucose.) Amylase is produced by the salivary glands, pancreas and small intestine.
- The enzyme **protease** digests protein molecules into amino acids. Protease is produced in the stomach, pancreas and small intestine.
- The enzyme **lipase** digests fat molecules into fatty acids and glycerol. Lipase is produced in the pancreas and small intestine.

It may help students to learn the enzyme names if they are told that starch contains amylose and fats are a kind of lipid. So the enzymes are named after the substance they work on, but with the ending 'ase'. Most digestion takes place in the small intestine (actually just the first 30 cm of it) due to enzymes secreted by the small intestine itself and by the pancreas. A common misconception is that the stomach acid helps digestion. It doesn't. The purpose of stomach acid is to kill bacteria, and in fact it hinders digestion by preventing most enzymes from working (which in fact is how the bacteria are killed). Only one particular protease enzyme called pepsin is produced in the stomach, and it is unusual in that it works best at pH 1. It might be worth mentioning to older students that we don't have a cellulase enzyme to digest cellulose (fibre), which is why cellulose cannot be absorbed and remains in the gut. Herbivores such as cows do have cellulase (thanks to the presence of symbiotic bacteria) and can digest cellulose.

The role of the liver in digestion is a bit more complicated than the other glands (pancreas and salivary glands). The liver doesn't produce any digestive enzymes but makes bile, which is stored in the gall bladder. Bile contains an alkali, which neutralises the stomach acid so that the pancreatic enzymes can work at their optimum pH (close to pH 7). Bile also contains bile salts, which act as a detergent, emulsifying large globules of insoluble fat into many tiny droplets. This emulsification increases the surface area of the fats enormously, allowing lipase enzymes to digest the fats much more quickly.

Digestion by saliva

A nice example of a digestive enzyme is the digestion of starch by amylase in saliva. This is a good introduction to enzymes, and it can also be used to reinforce the 'fair test' or controlled variables. Students drink a small amount of water from a plastic cup, swirl it around their mouth and spit it back into the cup. This saliva solution is then mixed with a starch solution and placed in a 40 °C water bath for 30 minutes. After this time the mixture is split into two test tubes so that it can be tested for starch (iodine test) and sugar (Benedict's test).

Eye protection should be worn when handling iodine. To ensure health and safety the plastic cups should be placed in a bowl of disinfectant before throwing away, and glassware should be similarly disinfected before normal washing.

An amylase solution can be used instead of saliva, but using students' own saliva is far more memorable.

This experiment can be used to practise data interpretation by adding control tubes, giving a results table like that in Table 2.1.

Table 2.1 Results of saliva digestion experiment

	Tube A (starch + saliva)	Tube B (saliva + water)	Tube C (starch + water)	Tube D (starch + boiled saliva)
starch test	✗	✗	✓	✓
sugar test	✓	✗	✗	✗

Students can be asked to explain the results for each tube. From tube A (the experimental tube) we may conclude that saliva digests starch to sugar, but how do we know that saliva doesn't contain sugar? Tube B rules out this possibility, and tube C is another control experiment to show that the result in tube A must be due to saliva and not due to some other feature of the procedure, such as the 40 °C water bath. Tube D shows that the saliva doesn't work if it is boiled (because the enzyme in it is destroyed or denatured). Only by comparing all four tubes can we conclude convincingly that an enzyme in saliva breaks down starch into sugar. This is a good example of the use of control experiments (as opposed to a controlled experiment, where confounding variables are controlled). The difference between controlled variables and control experiments is one of the most difficult concepts to understand. Students can also suggest variables that were controlled (temperature, time, equal volumes, etc.) and not controlled (enzyme concentration).

A simple version of the saliva experiment is to try chewing bread for as long as possible without swallowing – after 2 minutes it should start to taste sweet as sugar is digested from the starch in the bread. Students can do this at home.

Other investigations with digestive enzymes

There are many suitable investigations using digestive enzymes that can be carried out at this point. You need to guard against the danger that students will associate enzymes only with digestion and there is an argument for leaving the detailed study of enzymes to a separate topic, where enzymes in other roles can be studied. It's certainly worth mentioning that there are thousands of different enzymes in every living cell speeding up every reaction, not just digestion. These investigations are ideal for developing 'How Science Works' skills. Investigations using digestive enzymes include:

- Protease can be investigated using a suspension of 'Marvel' milk powder as a substrate and recording the time taken for the mixture to clear as the insoluble protein is digested to soluble amino acids. It has to be Marvel, as other milk powders contain little protein.
- Lipase can be investigated using full-fat milk as a substrate with a pH indicator (such as phenolphthalein) added to show the decrease in pH as fatty acids are produced.
- Amylase can be investigated by using a starch suspension as a substrate and using iodine to indicate the amount of starch remaining. Unfortunately iodine inhibits amylase so it can't be part of the reaction mixture. Instead, small samples can be taken at 1-minute intervals and added to iodine drops on a spotting tile. Alternatively iodine can be added to the reaction mixture at a certain time, and starch concentration measured by recording the intensity of the blue colour using a colorimeter.
- For older students the effect of temperature or pH on rate of reaction could be investigated for any of these enzymes. All the enzymes will have an optimum temperature of 40–60 °C (depending on the source of the enzyme). All the enzymes except pepsin will stop working at low pH, and this result can be linked to the purpose of stomach acid in killing bacteria and stopping most digestion.
- The effect of bile salts on the rate of digestion of fats by lipase can also be investigated. Bile salts are quite expensive and extremely smelly, and a few drops of clear detergent work just as well.
- Agar plates containing starch or milk can be used to assay amylase or protease enzymes respectively by placing samples of enzyme in small wells cut in the agar. Clear rings form around the wells as the enzyme diffuses through the agar, digesting the insoluble starch or protein into soluble products. The larger the clear ring, the more enzyme was present in the well, so this technique can be used to assay samples for the presence of digestive enzymes. A well containing water should be used as a control, and students could be asked to explain its purpose.

- Table jelly (but not agar jelly) is made with the protein gelatine, so it can be digested with protease. Make up a dark-coloured jelly, such as blackcurrant, in Petri dishes with a third of the recommended volume of water so that the jelly is firm. Cut wells with a cork borer and fill with samples of protease (e.g. biological washing liquid; contact lens cleaner; meat tenderiser; fresh pineapple juice; fig juice; pepsin; trypsin). Leave for a few hours and measure the increase in well diameter.

These 'jelly assays' are fun and simple for students to carry out, but it is important to realise that the size of the rings depends mainly on the rate of diffusion, not on the rate of reaction (which will usually be much faster). Imagine enzyme molecules randomly diffusing through the jelly matrix, slowly making progress outward from the well, but almost instantly reacting with any substrate molecules they meet on the way. So jelly assays are most useful for investigating the concentration of enzyme in the sample (which directly affects rate of diffusion and so size of the ring), but less suitable for investigating other factors such as pH or temperature, which affect rate of reaction and rate of diffusion.

■ Absorption

What happens to the small soluble molecules produced by digestion? They are absorbed through the wall of the small intestine into the bloodstream. Students don't generally realise that food that has been ingested and is in the gut is not technically inside your body. A simple diagram showing a human as a gut surrounded by a body can clarify this (Figure 2.5).

Figure 2.5
Absorption of food
in a human

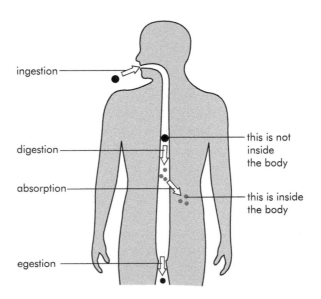

The contents of the gut are really (topologically speaking) part of the outside world! Food is not truly inside the body until it has crossed the gut wall and has entered body cells.

Absorption across the gut wall and into the blood takes place by diffusion and is a slow process. To speed it up the small intestine has a huge surface area (about half a football pitch in an adult). This surface area is achieved by the length of the intestine and by folding of the inner surface into villi and microvilli. These folds can be shown on a prepared slide of the small intestine, perhaps projected using a video microscope, using images from the internet or using a simple model of folded paper. Once again the link between structure and function can be emphasised.

Recall from the start of the topic that the purpose of eating is to obtain nutrients for growth and energy. So after absorption the nutrients are carried everywhere throughout the body by the blood and used for building materials and for releasing energy. The small molecules may be built up into large molecules again to make muscle, skin, bone, etc. using more enzymes. This process is called assimilation.

■ Egestion

Only undigested material is left in the gut, and it is egested through the anus as faeces. The undigested material is mainly plant cellulose cell walls (fibre), which we can't digest. Faeces also contain rubbed-off gut cells and bacteria from the large intestine. The presence of so many bacteria is why it is important to wash your hands after going to the toilet. A common misconception by younger students (sometimes reinforced by textbooks) is that food enters in the mouth and leaves by the anus. In fact the great majority of food leaves the gut in the small intestine and is absorbed into the blood. Only fibre cannot be used and passes out as faeces. Nonetheless, it is important to have fibre in the diet (i.e. from fruit and vegetables) so that peristalsis can work properly and solid faeces can be formed. Students often get confused with egestion and excretion. Egestion is the elimination of unabsorbed food from the gut (which has never really been in the body), while excretion is the elimination of waste material produced from within the body's cells. So faeces is egested while urine is excreted.

The fate of food can be summarised by a role-play exercise involving the whole class. Give every student a sticker so everyone can see what they are. The stickers will include glucose, amino acid, fatty acid, glycerol, water, some vitamins, some minerals, cellulose and the enzymes amylase, protease and lipase. The glucose, amino acid, fatty acid and glycerol students should join up to make starch, protein and fat molecules. Divide the room into two with a barrier of chairs. One side is the inside of the gut, the other side is the bloodstream and the chairs are the gut wall. Students are food molecules that have been ingested, so they start in the gut side of the room. Ask the students what they do next. The small molecules should be able to diffuse through the gut wall to the other side of the room. The starch, protein and fat molecules will be too big to cross the barrier, but can be digested by specific enzymes to make small molecules that can go through to the other side of the room. Only the cellulose will remain in the gut, and it will pass on to be egested as faeces.

Students can also describe what happens to their breakfast as a good summary of the whole topic. The description can be as detailed as you like, including the fates of the various components of the breakfast. Students should be encouraged to use all the technical terms they have learnt, such as the names of all the parts of the digestive system, the names of the enzymes and the words ingest, egest, digest and absorb.

2.3 Photosynthesis

Previous knowledge and experience

At age 11, students will probably have grown plants in pots and in gardens and will know that plants need water and light to grow, but they are unlikely to know much more about photosynthesis. Gardening advertisements can reinforce misconceptions about 'plant food'. Depending on teaching order, students may have studied respiration and know that all living cells respire.

A teaching sequence

The teaching sequence suggested here starts with the photosynthesis equation and uses that as a basis for various practicals investigating the production of the products (carbohydrate and oxygen) and uptake of the reactant carbon dioxide. Students are often famously unenthusiastic about plant biology, but they may be more enthusiastic about practical work. Most of the practicals described here can easily be carried out by the students themselves. Indeed, the science could be derived entirely from investigative work, with students deducing the equation themselves from their experimental results. The Science and Plants for Schools (SAPS) website has details of most of the investigations suggested here, as well as further background information on plants and photosynthesis.

■ Plant growth

A good way to test students' existing understanding of plant nutrition is to show them a plant seed and a fully-grown plant (the bigger the better), and ask, 'Where did most of the mass of the adult plant come from?' Most people will answer from soil or from water, or even from the Sun, but very few give the correct answer: from the air. It is very difficult for some students to believe that solid wood was built from thin air, and students may like to investigate the history of discoveries about plant nutrition, such as the experiments of Jean-Baptiste van Helmont (1577–1644; Figure 2.6), Jan Ingenhousz (1730–1799) and Joseph Priestly (1733–1804). Such research can add a human element to the subject and can result in a deeper understanding of the subject.

Figure 2.6
van Helmont's
experiment

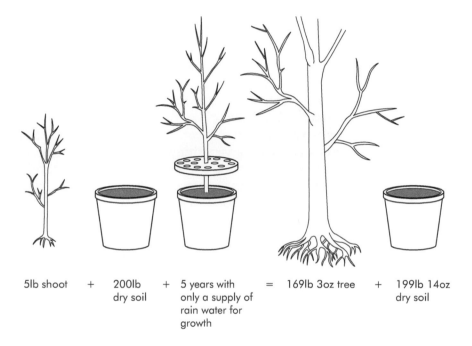

| 5lb shoot | + | 200lb dry soil | + | 5 years with only a supply of rain water for growth | = | 169lb 3oz tree | + | 199lb 14oz dry soil |

The discoveries of Priestly and others led to the equation for photosynthesis:

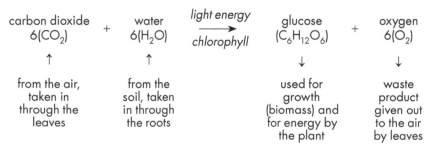

carbon dioxide $6(CO_2)$	+	water $6(H_2O)$	$\xrightarrow[chlorophyll]{light\ energy}$	glucose $(C_6H_{12}O_6)$	+	oxygen $6(O_2)$
↑		↑		↓		↓
from the air, taken in through the leaves		from the soil, taken in through the roots		used for growth (biomass) and for energy by the plant		waste product given out to the air by leaves

If students have already studied respiration then they should recognise this equation – it's the respiration equation backwards. Younger students should know the word equation while older students may need to know the symbol equation as well. It is fairly simple to balance the symbol equation, but this balancing is actually misleading. All the carbon and oxygen atoms in glucose come from carbon dioxide, while all the oxygen atoms in the product oxygen come from the water, and to show this correctly it is necessary to include water on both sides of the equation. This also means it's wrong to say that plants turn carbon dioxide into oxygen. It is more accurate to say plants turn carbon dioxide into glucose, and water into oxygen. The photosynthesis 'equation' is really a summary of some 30 separate steps, rather than a simple chemical reaction.

■ Leaves

Photosynthesis mostly takes place in leaves, which are well adapted to this job. Leaves are thin and flat to absorb as much light as possible, and they can turn to face the Sun. Time-lapse videos of this heliotropism (which was first recorded by Leonardo da Vinci) can be found on YouTube or the BBC video *The Private Life of Plants*. A transverse section of a leaf can be studied under the microscope using prepared slides or very thin sections of a leaf clamped between two halves of a carrot and cut with a sharp blade. Stomata can also be observed under the microscope by peeling the lower epidermis off a suitable leaf (red hot poker, *Kniphofia*, rhubarb stem and elephant's ear saxifrage, *Bergenia*, work well) or peeling dried nail varnish that has been painted on to the lower epidermis. Suggested methods are given on the SAPS website. This could be a good time to revise parts of plant cells, in particular how palisade cells are adapted to their function of photosynthesis. Older students can use a graticule to measure the density of stomata on a leaf surface and investigate how this stomatal density varies in plants from different environments.

■ Chlorophyll

Photosynthesis depends on chlorophyll, a green compound containing magnesium found inside chloroplasts. Chlorophyll absorbs light energy, using it to split water and finally storing it as chemical energy in glucose. Students can easily extract chlorophyll from leaves by grinding some fresh spinach leaves in propanone and filtering, or by scraping the juice out of grass leaves on a slide (protocol on SAPS website). Water should be removed by evaporating the solvent using a hairdryer and re-dissolving in fresh propanone.

The different pigments present in this leaf extract can be separated using chromatography. Good separation can be achieved in just 5 minutes using thin layer chromatography (Tomkins and Miller, 1994). It should be possible to see chlorophyll a (dark green), chlorophyll b (pale green), carotene (yellow) and leutin (brown). Different coloured leaves, or red or brown algae, can also be used to obtain different mixtures of photosynthetic pigments. The chlorophyll solution can also be used to investigate the absorption of light. An intense white light source can be split into a spectrum using a glass prism in a dark laboratory and a cuvette of chlorophyll inserted into the light path (Figure 2.7). The chlorophyll absorbs the red and blue parts of the spectrum, but transmits the green and yellow parts. This is why plants look green. It means that photosynthesis uses red and blue light, but not green light – students often get this the wrong way round.

Figure 2.7 Absorption of light by chlorophyll

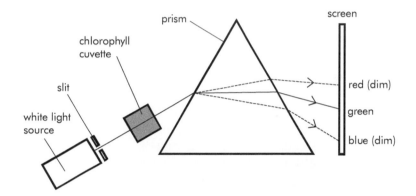

■ Demonstrating photosynthesis

Many data-logging systems include oxygen and carbon dioxide gas sensors (Delpech, 2006). These sensors can be used to demonstrate the changes in the gases due to photosynthesis in leaves and small plants (Figure 2.8).

Figure 2.8 Using gas sensors to demonstrate photosynthesis

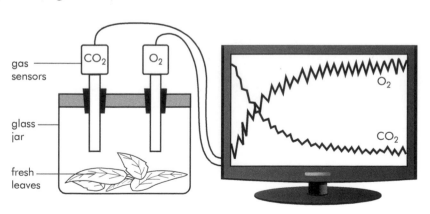

Changes in both gases can be observed in real time as conditions, such as light intensity, are changed. The gas sensors have the advantage that familiar land plants can be used, both oxygen and carbon dioxide can be monitored simultaneously and the changes can be projected on to a screen. However, the expense means it is unlikely that they can be used for individual or small group investigations. Gas sensors are therefore best used as a demonstration, perhaps as a stimulus at the beginning of the topic.

■ Investigating the factors required for photosynthesis

Students can test for the main product of photosynthesis: carbohydrate. It's not easy to test for glucose, but we can easily test for starch (which is made from glucose in leaves) using the iodine test. This investigation can be done using any laboratory plants, such as *Pelargonium*. The plants need to be de-starched first by keeping them in a dark cupboard for 2 days, so that all the starch is used up in respiration. Then the plants can be set up to test for various factors such as light (by covering a leaf in foil, perhaps with a hole cut in it); carbon dioxide (by placing the leaf in a flask containing potassium hydroxide to absorb the carbon dioxide); stomata (by painting one or both surfaces with petroleum jelly) and chlorophyll (by using a variegated leaf). The plants are exposed to bright light for 24 hours (Figure 2.9a).

Figure 2.9
Investigating the factors required for photosynthesis. a Setting up the experimental plant. b Testing the leaves for starch

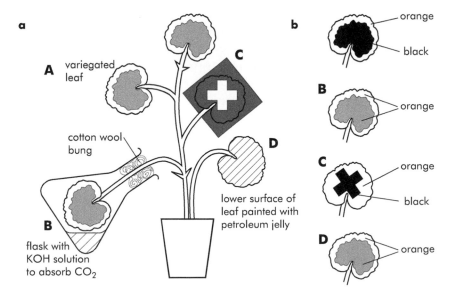

PROCEDURE

After 24 hours students test each leaf for starch. The test can be done on whole leaves, or on small disks cut from leaves with a cork borer pressed against a rubber bung. There are four steps in the starch test:

1 Place a leaf in a 90 °C water bath for 1 minute to kill the leaf cells.
2 Place the leaf in a boiling tube of ethanol and place the tube into the 90 °C water bath for 5 minutes while the ethanol boils. All the green chlorophyll should dissolve in the boiling ethanol and the leaf should be decolourised.
3 Important: no naked flames should be used while using ethanol. Eye protection is needed.
4 Remove the leaf with forceps and wash in cold water.
5 Place the leaf in a Petri dish or on a tile and flood with iodine solution. Draw or photograph the appearance of the leaf.

Typical results are shown in Figure 2.9b. Students should make their own conclusions about what factors are needed for photosynthesis. They often find this difficult, focusing (understandably) on what factors were present, e.g. 'This shows that water is needed for photosynthesis as the plant was well-watered and it made starch'. In fact the interesting conclusions are made by observing what happens when the factor is absent, e.g. 'This shows that carbon dioxide is needed for photosynthesis as leaf B did not have carbon dioxide and did not make starch'.

■ Measuring the rate of photosynthesis

We can also measure the production of the other product of photosynthesis – oxygen gas. Unlike the starch test, it is fairly easy to measure the rate at which oxygen is produced, and so investigate factors that affect the rate of photosynthesis.

The most common way to do this is to use a pondweed that grows under water. A length of pondweed is placed upside down in 0.25 M (2%) potassium hydrogencarbonate solution and the oxygen produced by photosynthesis appears as bubbles from the cut end. The simplest protocol is to weight the pondweed with a paperclip or Plasticine so that it sinks to the bottom of the tube and the oxygen bubbles can be seen rising through the water (Figure 2.10a). The number of bubbles rising from the pondweed in a minute can simply be counted and used as a measure of the rate of photosynthesis. A more sophisticated apparatus, called a photosynthometer or Audus apparatus, collects the bubbles in a capillary tube where the length (and therefore the volume) of the gas produced in a set time can be measured (Figure 2.10b). The gas can also be collected in an upturned measuring cylinder over a funnel and volume measured that way (Figure 2.10c). The gas in the cylinder can be tested with a glowing splint to demonstrate that it is oxygen (although this doesn't always work as the gas is by no means pure oxygen). The oxygen released can also be measured using a dissolved oxygen probe connected to an oxygen meter or data logger. This method is quantitative but dissolved oxygen probes can be tricky to set up.

Figure 2.10
Measuring the rate
of oxygen
production by
pondweed.
a Counting
bubbles;
b Measuring the
volume using a
photosynthometer;
c Collecting the
gas in a measuring
cylinder

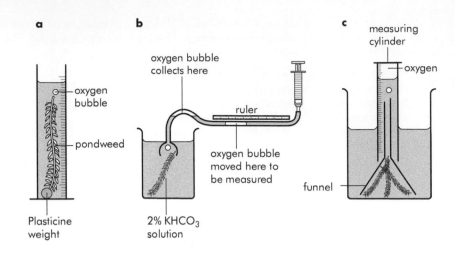

Pondweed experiments have a reputation for being unreliable, but in fact work very successfully as class experiments if precautions are taken. The tropical pondweed species *Cabomba* or *Lagarosiphon* have been found to work better than the more common temperate *Elodea* species (Eldridge, 2005). They can all be found in pond and aquarium shops and from suppliers of biological materials. The plants should be kept in large sunny aquarium tanks and short lengths cut cleanly with sharp scissors underwater immediately before use. The 2% potassium hydrogencarbonate solution is needed to provide plenty of carbon dioxide for photosynthesis, and the potassium salt is reported to be more reliable than the sodium salt. Normal room lights and desk lamps are far too dim for photosynthesis, and a very bright light source is needed to encourage a good rate of bubbling. 150 W halogen lamps (used for outdoor illumination) give out plenty of light, but they also give out lots of heat, so a heat filter, such as a 1 litre 'medical flat' bottle filled with water, is needed. Fluorescent tubes are cool light sources so they don't need a heat filter, and 'daylight' tubes are ideal for photosynthesis since they give more blue light to mimic natural daylight. Compact fluorescent tubes (energy-saving bulbs) can be used in normal light fittings, while larger light banks with several fluorescent tubes, suitable for photosynthesis experiments as well as for growing plants generally in the laboratory, can be obtained from Blades Biological. Hydroponics suppliers such as Progrow sell grow lamps with an output spectrum ideally suited for photosynthesis, but these lamps are expensive. For a controlled experiment the pondweed should be shielded from other light sources by placing it in a cardboard box or in a tent made with a coat over a lab stool.

Various different independent variables can be investigated using any of these techniques. The light intensity can be changed by inserting neutral density filters in front of the lamp, or by simply moving the lamp further away. The light intensity varies as $1/d^2$, but it is probably better to measure the intensity with a lux meter. The colour of the light can be changed by placing coloured filters in front of the lamp, although care has to be taken that this doesn't affect the intensity as well. The carbon dioxide concentration can be altered by changing the amount of potassium hydrogencarbonate added to the solution. The temperature can be changed by placing the pondweed tube in a glass water bath, such as a large beaker. All these investigations can be planned by students themselves, choosing which variables to control and how to quantify the outcome.

Another way of using the oxygen produced in photosynthesis is to use floating leaf disks. Small disks are cut from leaves and placed in a potassium hydrogencarbonate solution. The oxygen produced by photosynthesis in the leaf disks causes the disks to rise to the surface and float, and the time taken to rise can be used as a measure of the rate of photosynthesis (details on SAPS website).

Simulations of the pondweed experiments can be found on the internet (see resources at end of chapter). These simulations can be used as extensions to laboratory experiments, but they are also valuable as evaluation exercises. Are the simulations realistic? What assumptions have been made? How are the simulations different? Are there any mistakes? This sort of evaluation can really test understanding of the process of photosynthesis.

■ Limiting factors

Experiments such as the ones above show that the rate of photosynthesis depends on a number of factors, including light intensity, temperature and carbon dioxide concentration. But at any given time there can only be one factor that is actually controlling the rate – the limiting factor. This is the factor that is in shortest supply, or furthest from its optimum. An analogy can be made with a chain that is only as strong as its weakest link. Limiting factors crop up in many areas of biology, but are often tackled in the context of photosynthesis. Students should be able to interpret graphs such as the one shown in Figure 2.11.

Also note that in plant production in greenhouses, carbon dioxide is added to increase photosynthesis and, therefore, crop yield.

Figure 2.11 Graph used to identify limiting factors. At low light intensities (A) the rate of photosynthesis increases as the light intensity increases, so light must be the limiting factor. At higher light intensities (B) the rate of photosynthesis stays the same even if the light intensity increases. This means that light is not the limiting factor, and the rate of photosynthesis must be limited by some other factor

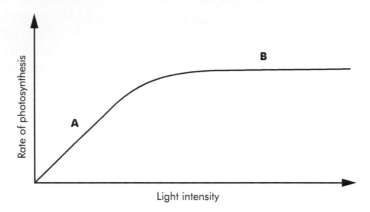

■ Photosynthesis and respiration

Plants are amazingly self-sufficient. They use the glucose they make in photosynthesis as the basis for making everything else in their cells, i.e. their biomass. This includes: starch for storage in roots and leaves; fats for energy store and for cell membranes; cellulose to make cell walls; sucrose for transport around the plant; and proteins for growth. For some of these compounds plants need other elements, such as nitrogen and sulphur. They get these two elements in the form of the minerals nitrate and sulphate, which they take up from the soil by active transport through their roots. Students should be encouraged to use the word 'minerals' rather than 'nutrients' or 'food' for these inorganic substances absorbed from the soil. Bottles of 'plant food' from the garden centre contain minerals.

Plants also use the glucose for respiration to release energy, so the glucose is turned straight back into carbon dioxide and water. All living cells respire all the time, including plant cells. So in the dark plants give out carbon dioxide and in the light plants give out and take in carbon dioxide at the same time, but overall take in far more carbon dioxide, since they have a much faster rate of photosynthesis than of respiration. In fact, plants take in enough carbon dioxide to compensate for the carbon dioxide given out by all the animals, plants, fungi and microbes on Earth. It may be appropriate at this time to link this delicate balance of carbon dioxide to global warming, which is due to a global increase in respiration and burning and a global decrease in photosynthesis as a result of human activities.

It may also be appropriate to consider that the plant biomass, manufactured almost entirely from carbon dioxide, is the basis of all food chains and is the food for us and our farm animals. This idea can be linked to the carbon cycle (see Chapter 9).

■ Investigating changes in carbon dioxide

Students can investigate changes in carbon dioxide concentration to show that carbon dioxide is taken up by plants in photosynthesis and given out by plants in respiration. The most common way to do this is to use hydrogencarbonate indicator solution. Since carbon dioxide forms a weak acid in solution, its concentration can be detected using this very sensitive pH indicator. Figure 2.12 shows the colour changes.

Figure 2.12 Colour scale for hydrogencarbonate indicator

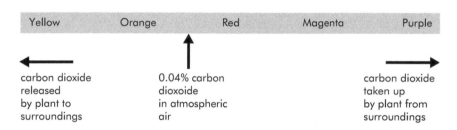

A typical experiment uses pondweed again and might have three boiling tubes as shown in Figure 2.13.

After a day in the light, tube A stays red, tube B turns purple as carbon dioxide is taken up by the plant for photosynthesis and tube C turns yellow as carbon dioxide is released from the plant by respiration. Students should be able to explain the colour changes in terms of photosynthesis and respiration.

A different and very successful way to do the hydrogencarbonate experiment is to use algae instead of pondweed (Eldridge, 2004). The algae are grown in bottles prior to the experiment and then entrapped in sodium alginate beads to form 'algal balls', which can be used in place of pondweed. The algal balls have the advantage of being quantifiable, easy to handle and quick to respond to changes in conditions. Changing light intensity gives a range of colours of hydrogencarbonate indicator, which can be quantified in a colorimeter and used as a measure of the rate of photosynthesis. The light intensity that gives no change in colour of the indicator represents the compensation point, where the rate of photosynthesis is equal to the rate of respiration, so there is no net change in carbon dioxide.

Figure 2.13
Investigating
changes in carbon
dioxide
concentration using
hydrogencarbonate
indicator

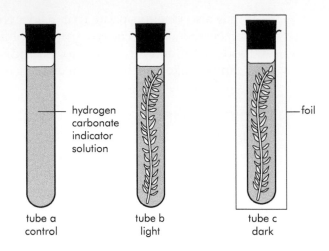

hydrogen
carbonate
indicator
solution

foil

tube a
control

tube b
light

tube c
dark

Other resources

Suppliers

Blades Biological (www.blades-bio.co.uk) – for pondweed, light banks, algae and growth media

Sciento (www.sciento.co.uk) – for algae and growth media

Progrow (www.progrow.co.uk) – for hydroponics grow lights

Lee Filters (www.leefilters.com) – for light filters

Generic websites

Mackean's Biology Resources: www.biology-resources.com

Practical Biology: www.practicalbiology.org/

Skoool: www.skoool.co.uk/

Wikipedia: http://en.wikipedia.org

Websites related to diet and digestion

The UK Food Standards Agency: www.eatwell.gov.uk

The British Nutrition Foundation: www.nutrition.org.uk

Diet analysis software, from Multimedia Science School: www.science-school.co.uk

Teachers TV pig's gut dissection: www.teachers.tv/videos/ks3-ks4-science-demonstrating-biology-it-takes-guts

The OBs, from Wellcome: www.wellcome.ac.uk/Education-resources/Teaching-and-education/big-picture/all-issues/obesity/student-activity/index.html

ABPI resource library. Search for digestion: www.abpischools.org.uk/page/resourcelibrary.cfm

Dying to be Thin, from Nova: www.pbs.org/wgbh/nova/thin/

ADAM Medical Animation Library: www.pennmedicine.org/health_info/animationplayer/

Digestive system animation, from John Kitses: http://kitses.com/animation/swfs/digestion.swf

Your digestive system, from Kids Health: http://kidshealth.org/kid/htbw/digestive_system.html

Digestive system, from Discovery Kids: http://yucky.discovery.com/noflash/body/pg000126.html

Balanced diet, from NGFL-Cymru: www.ngfl-cymru.org.uk/vtc/balanced_diet/eng/Introduction/default.htm

The digestive system, from NGFL-Cymru. Very nice. Includes enzymes. www.ngfl-cymru.org.uk/vtc/digestive_system/eng/Introduction/default.htm

BMI, from Centers for Disease Control and Prevention: www.cdc.gov/healthyweight/assessing/bmi/index.html

Cholesterol animation, from healthScout: www.healthscout.com/nav/animation/1/main.html

Websites related to photosynthesis

SAPS (Science and Plants in Schools): www-saps.plantsci.cam.ac.uk/

Pondweed simulation, from Cambridge Assessment (free): www.cambridgeassessment.org.uk/ca/Spotlight/Detail?tag=Enigma

Pondweed simulation, from Explore Learning (paid for): www.explorelearning.com/index.cfm?method=cResource.dspDetail&ResourceID=395

The Tomato Zone: www.thetomatozone.co.uk/

UK Agriculture: www.ukagriculture.com

References

Delpech, R. (2006). Making the invisible visible: monitoring levels of gaseous carbon dioxide in the field and classroom. *School Science Review*, **87** (320), p. 41.

Eldridge, D. (2004). A novel approach to photosynthesis practicals. *School Science Review*, **85** (312), pp. 37–45.

Eldridge, D. (2005). Cabomba – a reliable alternative to *Elodea*? *SSERC Bulletin*, **215**, pp. 10–12.

Tomkins, S.P. and Miller, M.B. (1994). A rapid extraction and fast separation of leaf pigments using thin layer chromatography. *School Science Review*, **75** (273), pp. 69–72.

3 Transport in organisms

Mark Winterbottom

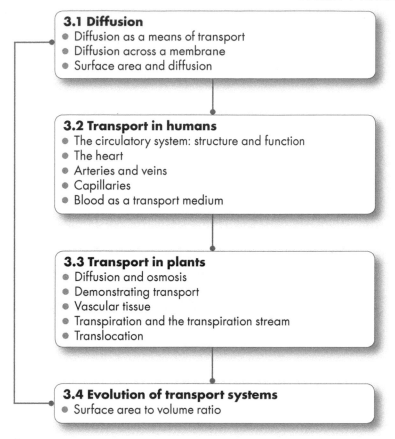

3.1 Diffusion
- Diffusion as a means of transport
- Diffusion across a membrane
- Surface area and diffusion

3.2 Transport in humans
- The circulatory system: structure and function
- The heart
- Arteries and veins
- Capillaries
- Blood as a transport medium

3.3 Transport in plants
- Diffusion and osmosis
- Demonstrating transport
- Vascular tissue
- Transpiration and the transpiration stream
- Translocation

3.4 Evolution of transport systems
- Surface area to volume ratio

Choosing a route

If you introduce this topic by saying, 'Today we're looking at transport', you'll get a lot of blank looks! Lorries, ships and planes might spring into some students' minds. Other students may focus on transport of electrical impulses (covered in Chapter 5) or transport of food through the digestive system (covered in Chapter 2). However, this chapter does not deal with any of these ideas, but focuses on the transport systems themselves. In animals, students are usually happy with the role of the blood in moving the reactants and products of respiration around the body, and carrying materials between organs. In plants, students realise that all parts of the plant need water and minerals, but their understanding of the mechanisms of transport can depend so much on their understanding of osmosis that you can easily trip up when trying to support their learning. So what's a good approach?

To understand transport in animals and plants, students need to understand diffusion. So let's start there. You've then got a choice; you could help children to understand the need for a transport system and why such systems have evolved, although many teachers introduce transport in the context of its purpose: in humans to supply cells with raw materials for respiration, to carry waste products away from cells, and to transport materials between organs. Hence, given that transport is most obvious in animals, you may want to move straight on to look at how transport works in humans, and return to the evolution of transport systems at the end of the topic. Having understood some of the principles of transport via the human circulatory system, you can extend students' understanding to transport in plants.

3.1 Diffusion

Previous knowledge and experience

Students may have met the idea that smells spread through the air, and that solutes will spread out through a solvent. They may be less likely to understand the particle model of diffusion, and that solutes can diffuse across a membrane. Some may know the definition of diffusion (the net movement of particles from a region of high concentration to a region of low concentration), but may not understand the idea of 'net' movement. Living things have a number of adaptations to increase surface area, often involving folding of membranes, which maximise rates of diffusion across the membrane. Students can find the effect of these difficult to understand.

A teaching sequence

■ Diffusion as a means of transport

One way to start this series of lessons is to get the students to look at some pond water or cultures of unicellular organisms or simple filamentous algae. Ask the students to work in pairs, to consider the following questions while they are looking at the organisms and to write down their thoughts as they go along:

- What do organisms take from their environment?
- What wastes do organisms release into their immediate surroundings?
- How do materials move into and out of cells?

Most students find looking at these organisms with a low-power microscope fascinating and get quite excited by the experience. Higher attaining students are likely to suggest that diffusion is involved in the exchanges between cells and their immediate environment; while others may simply say that materials pass in and out of the cells. It is important at this stage to reinforce the particle model of diffusion. You can use some simple demonstrations and models to help. First, reinforce the idea of diffusion as a slow process which involves spreading out:

- Observe diffusing potassium manganate(VII) in a beaker of water on an overhead projector (OHP).
- Ask students to undertake a thought experiment about someone in their family cooking and why they can smell the aroma while lying in bed.

Then challenge students' misconceptions about how diffusion works. Without giving them any more information, cluster a group of students in a corner (equivalent to the potassium manganate(VII) just dropped into the water) and ask them to role-play diffusion. Usually, they all walk away from their original location, leaving a space with no particles. Discuss why this is wrong, and use the discussion to counter the intuitive idea that particles *want* to go to the region of low concentration. To reinforce the random nature of the movement of particles, you could try some of the models below:

- Diffusing black and white beans or balls in a tray. Place the black and white beans/balls at opposite ends, and gently shake the tray – eventually they will spread out and mix. Ask them to watch an individual bean/ball, and notice that it moves randomly in all directions, even though there is net movement, which makes the balls mix.
- ICT simulation of diffusion. Try to get one with variable speed control, or which moves slowly so students can follow individual particles and fully realise the random nature of movement.

You can assess students' learning by asking them to predict and explain what happens in different models of diffusion.

■ Diffusion across a membrane

If you decided to cover evolution of transport systems first, ask students to think back to the microorganisms they met there. Remind students about the fact these are membrane-bound, and through discussion, help them to realise that diffusion must be able to happen through a membrane. If you haven't met the microorganisms with students yet, simply show them some videos of amoeba (you can find some on YouTube), and go through the same ideas. Drawing on their understanding of how diffusion works, in pairs, ask them to make hypotheses about properties of membranes, which are needed to allow diffusion to happen. As a class, arrive at a functional model of a membrane as a surface that has holes in, which will allow particles to pass through. You could use a colander or sieve as a useful analogy.

You may want to use a simulation that shows diffusion across a membrane, or include a cardboard membrane in one of the models above to reinforce the idea. If you do use a simulation, be careful to avoid any talk about diffusion of water; that comes later! As you work through the simulation, ask students to predict how the particles will move. You could also use an animation that doesn't represent the particle model of diffusion particularly well, and ask students to identify the flaws in the model. One example can be found here: http://student.ccbcmd.edu/courses/bio141/lecguide/

unit1/prostruct/passive_flash.html. If students have come across, or use, the term 'osmosis' help them appreciate that osmosis is a feature of all organisms not just plants.

■ Surface area and diffusion

The transport systems of animals and plants have adaptations that help to maximise the rate of diffusion across a membrane, often involving folds in the membrane. If you think about a membrane as a surface with holes in, particles will diffuse through these holes. If you can fit more holes in the same length by making the membrane fold (hence increasing the area of membrane and the number of holes), this will increase the overall rate of diffusion. This can be hard to understand 'on the hoof', and it is worth coming back to this explanation wherever relevant below. To assess students' understanding, you could ask students to storyboard an animation that shows the effect of a folded membrane on the rate of diffusion. If students work in groups to devise the storyboard, it would probably help to support the understanding of less advanced students.

3.2 Transport in humans

Previous knowledge and experience

Students should have already met the heart as an organ and perhaps blood as a tissue. Most students will recognise that the food they take in through their mouths and the oxygen in the air they breathe must get around their body somehow. They are likely to know that the heart functions as a muscular pump, which circulates blood to all parts of the body. They will have met the idea that blood flows around the body in blood vessels. Students are also likely to have some idea about the effects of different lifestyles on the health of the heart.

A teaching sequence

Given that students will be familiar with organ systems, it is sensible to introduce the circulatory system as a system within which blood (as a transport medium) is pumped around the body. Focus on the way in which the system functions as a whole (as a double circulation: one circulation to the lungs and a separate one to the body, with the heart operating almost as two separate pumps), and then focus in, in more detail. Look at the structure and function of the heart, to appreciate how it operates as a pump within the double circulation. Then look at the ways in which the arteries and veins are adapted for their functions. Finally, look at the structure and function of capillaries in allowing materials to pass in and out of the blood. This is a good point at which to look at the constituents of the blood and their role in transporting oxygen and glucose to cells, and taking away carbon dioxide and water.

You will face a number of common misconceptions in this topic. Students can think of the heart as a pulsating bag (a single pump), and find it difficult to visualise the double circulation; often they imagine that blood travels from the heart, through the lungs and directly to the body. Many students will think that arteries always carry oxygenated blood, while veins only carry deoxygenated blood. In fact, direction of flow relative to the heart is the distinguishing feature between arteries and veins (the pulmonary artery actually carries deoxygenated blood and the pulmonary vein carries oxygenated blood). Some students may mistakenly think that oxygen is carried in the blood plasma, and some students may think that deoxygenated blood is blue (because it is often depicted that way in textbooks).

■ The circulatory system: structure and function

Understanding the need for a double circulation can be difficult. To ensure students are familiar with the structure, you may want to give them a diagram of the human circulatory system (Figure 3.1) and ask them to suggest what happens to the level of oxygen and carbon dioxide in the blood as it passes around the two circuits. Then, in pairs, give them the equivalent diagrams for fish (single circulation), amphibians, birds and reptiles (incompletely separated double circulation) and ask them to list any differences they can see (compared to the human circulation) and to propose any consequences which may arise from those differences. For less advanced students, try to focus their attention on whether oxygenated and deoxygenated blood mix together, and the pressure at which blood can be pumped. Put pairs together into a foursome and ask them to summarise the benefits of a double circulation.

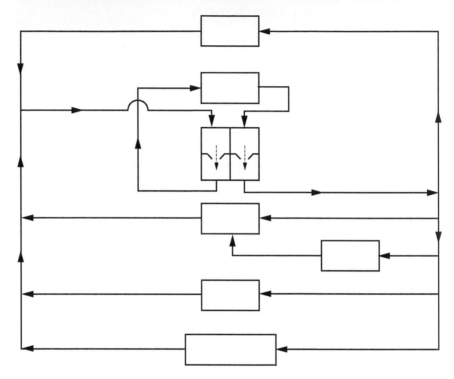

Figure 3.1 A blank flowchart of the circulatory system

■ The heart

Dissection can be an emotive subject and one about which a school science/biology department will have a policy. You may wish to do a dissection alone and take photos for the class to see. If possible, though, real material is the most interesting for students to see. Depending on your class, you could do a demonstration dissection of the heart (with a webcam pointing down on to the dissection, and projecting images on to the whiteboard), or you could let students dissect hearts in small groups. Some students may object to dissection and so a valid alternative would need to be provided for them (virtual dissections on the internet may be suitable, or they may be happy to watch the webcam images on the screen). Some students may feel nauseous and some could even faint – ensure they are all sitting on chairs before dissecting. You can obtain hearts from supermarkets (although they usually are trimmed up too neatly) or from a butcher (where you can ask for the main veins and arteries to be left protruding from the heart).

The dissection should establish the features shown in Figure 3.2, unlabelled copies of which could be supplied to the students. You can find protocols on the internet to help you dissect well enough to yield the 'textbook' picture (see websites listed at the end of the chapter). However, don't forget to look at the outside of the heart first, and don't be afraid to poke your fingers down tubes before and during dissection to see where they lead to! You can even mimic the action of the heart by filling it with water through the arteries that come out of the top, and squeezing the base to force the water back out. The first cut you make should remove the bottom 2 cm from the base of the heart. By looking at the cut end, you can see the distinction between the wall thickness of the left ventricle (which pumps blood to the whole body) and right ventricle (which only pumps blood to the lungs), providing a useful link back to the double circulation. If students do their own dissection, ask them to label their dissected structure with pins to show the direction taken by the blood through the heart. If you project the dissected structure on to a whiteboard, ask students to draw in the route taken by the blood.

Figure 3.2
Dissected heart

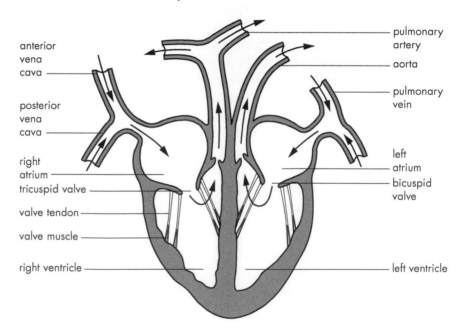

To ensure students understand how blood is moved through the heart, ask them to find an animation of the cardiac cycle on the internet, and write a commentary for it, which should indicate at each stage which muscles are contracting, which valves are open and closed, and in which direction the blood is flowing. Students should listen to some of the commentaries as a class, giving each other feedback; then, as a follow up exercise, you could ask them to create a flicker book to summarise the process. For less advanced students you could provide pictures and descriptions of each stage of the cardiac cycle for them to match up.

■ Arteries and veins

At this stage it is sometimes helpful to ask students to write a design brief for the vessels that will carry blood away from the heart, and a design brief for the vessels that will bring blood back towards the heart. This is a really good opportunity to use a snowball technique, where students start working on their own, then in pairs, and then in fours, gaining new ideas at each transition. Given that this is quite a challenging task, using group work in this way also supports differentiation for some students. Its aim is to focus students on what adaptations each type of vessel requires.

At this point it can be useful for students to examine prepared slides of transverse sections of arteries and veins under the microscope (many biology departments have a store of prepared slides). Ensure you spend some time explaining that a transverse section has cut *across* the vessel, and that they are looking at the cut end through the microscope (it can be difficult for students to relate two-dimensional images to three-dimensional structures). Projected images of the same sections using photomicrographs would be helpful to make sure that students are actually focused on what you want them to see. You can find good examples by using the search terms 'transverse section vein' on an internet search engine.

Ask students to draw what they see and use their drawings, or prepared diagrams, to build models of the blood vessels (you could ask them to bring in appropriate materials). Students should complete a table like Table 3.1 (with a description of the structure and its function), and then evaluate each other's models. Hopefully, students should recognise that blood vessels bringing blood to the heart (veins) differ in structure from those taking blood away from the heart (arteries) and that the differences reflect the differences in blood pressures that they have to accommodate. You can test their understanding of the adaptations of the vessels, by asking, 'What would happen if an artery had the adaptations of a vein?' and 'Why can you feel your pulse?'

Table 3.1 Table to show the differences between arteries and veins

Feature	Artery	Vein
Relative thickness of wall		
Amount of elastic tissue, including muscle		
Relative size of lumen		
Valves	No valves present except at the base of the aorta and pulmonary artery	Pocket valves present

The reason we can feel our pulse is because of the recoiling action of the artery wall as the heart pumps blood through. One novel way of demonstrating the pulse is to attach a drawing pin to the base of a safety match. If the drawing pin is delicately balanced over the radial pulse with the arm resting firmly on a flat surface, it is possible to actually see the pulsating action of the left ventricle (Figure 3.3).

Figure 3.3 Using a drawing pin and matchstick to demonstrate the action of the pulse

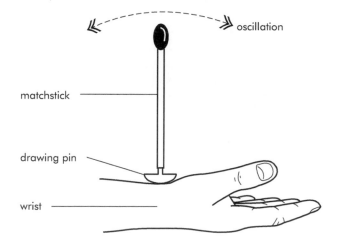

oscillation

matchstick

drawing pin

wrist

■ Capillaries

Having established how blood is moved around the system, in a particular direction, the next step is to understand how things 'get on and off' the transport system. Students know that arteries and veins work as tubes carrying blood from one place to another, and to do so efficiently they presumably must not leak.

However, if molecules are going to enter and leave the blood, there must be a third type of vessel that does allow molecules to enter and leave. This is the capillary. It is possible to see capillaries by placing a drop of cedarwood oil on the skin of one finger, just below the cuticle of the fingernail. By looking at the skin under a stereo-microscope, the tiny threads of surface capillaries are visible.

Capillaries not only let things in and out; capillary beds provide a much greater surface area for exchange of materials than a single artery would have done (even if it had a very thin wall). This is fairly obvious (if you split one tube into many smaller tubes, the membrane area is increased). However, you could use a model to help make it clear. The capillaries are in fact so narrow that the red blood cells have to squeeze through, pushing up against the capillary wall and minimising the diffusion distance for solutes.

■ Blood as a transport medium

Discussion about materials entering and leaving the transport system in capillary beds provides a natural progression to the nature of blood and the way it functions as a transport medium. Always take advice from the head of biology or head of science before engaging in any practical where you intend to use your own or your students' blood. You may be surprised at what is possible, but unless you take advice at an early stage you could easily put your own health or that of your students at risk. CLEAPSS can also provide additional advice.

The first step is to look at what's in blood, and then to look at how it transports materials. It can be difficult to persuade some students that blood isn't just a liquid. If you order some lamb's blood from your local butcher and centrifuge it, it's possible to demonstrate that there is a yellow fluid (plasma) along with some thicker material (the blood cells). You can find some useful animations of red blood cells on YouTube, and your department may have some prepared slides of blood that students can look at under the microscope, to establish that the 'non-plasma' part of the blood is made of red blood cells, white blood cells and small fragments of cells called platelets.

At this point you could ask students to work in groups, listing the things that must be transported around the body in the blood. Invite them to record where these things come from, where they are going to, and what they are needed for/produced by. Having spent 2 minutes setting out their ideas in a table, they could swap sheets with another group who then add to their ideas or correct them.

Set up a game of 'Blind Date', where three of the class have role cards describing their adaptations (as white blood cell, platelet or red blood cell). In terms of transport, the white blood cells and platelets have no role. However, the red blood cells are adapted to carry oxygen. The class has to 'choose' which of the Blind Date participants is involved in oxygen carriage. To reinforce the ideas on their role card, ask students to produce a job advertisement in groups, advertising the role of 'oxygen transporter', and requiring applicants who have the appropriate adaptations (e.g. a biconcave shape, the lack of a nucleus, a thin and permeable membrane, flexibility and the presence of haemoglobin).

To help students understand how the biconcave shape increases surface area, you could ask them to make a model of a red blood cell by cutting into a conventional bathroom sponge, and then measure its surface area by sticking squared paper to the sponge and measuring the total area of the paper. By comparing this to an intact sponge of the same width, students can demonstrate to themselves that biconcave cells have more surface area and therefore can absorb and lose oxygen more easily. As the difference between the surface area of a biconcave sponge and a normal sponge is not large, you may need to collate data from each student to find a class average in order to make the difference clear.

To help students understand why haemoglobin takes up or loses oxygen in different conditions, introduce the reaction as an equilibrium, in which oxyhaemoglobin is on one side, and haemoglobin and oxygen on the other. Where there is lots of oxygen (e.g. in the lungs) the equilibrium moves towards oxyhaemoglobin. Where there is little oxygen (e.g. around the body's cells) the equilibrium shifts to release more oxygen, which then diffuses out of the red blood cell, into the plasma, and out of the capillary into the tissue fluid surrounding the cells. If you can obtain lamb's blood you can demonstrate this. As soon as you get it, add $5\,cm^3$ of 0.1% sodium oxalate per litre of blood to prevent clotting. Place equal amounts in three flasks and bubble oxygen through one, carbon dioxide through another and leave the other as a control. The blood in the oxygen flask will turn red, showing that haemoglobin binds to oxygen in high oxygen conditions. The blood in the carbon dioxide flask will turn dull red/purple, showing oxygen being released from haemoglobin in high carbon dioxide conditions.

Finally, give students a series of bullet points that describe how materials leave and enter the capillaries, to and from the body cells bathed in tissue fluid. Students have to convert the bullet points into a diagram of exchange occurring at a capillary bed. Use this activity to highlight the idea that glucose and carbon dioxide are transported in solution, in the plasma.

■ Lymph as a transport medium

Although not commonly included in specifications at GCSE level, the lymphatic system operates alongside the blood system. Some students will be aware of swollen lymphatic glands when they are ill, or they may be aware of the role of lymph in the transport of digested fats in the digestive system. Given that little detail is usually required, it is commonly adequate, and indeed interesting, for students to identify similarities and differences between the blood system and the lymphatic system, both in terms of structure and function.

Further activities

- To bring together their learning so far, ask students to create a board game that shows how the transport system works in humans.

Enhancement ideas

- There are numerous opportunities to link social, moral and ethical issues to students' work in this topic, and your teaching of any of the ideas above could be set in any of the following contexts:

  - valve by-pass operations
  - heart transplants
  - use of pigs' hearts as a means of treating cardiac disease in humans
  - the ethics of developing transgenic organisms to provide organ banks
  - blood donation/transfusion
  - blood transfusion and HIV.

3.3 Transport in plants

Previous knowledge and experience

Students are likely to have met the idea that plants absorb water from the soil via their roots and that water passes up the stem to the leaves. They are also likely to be aware that plants photosynthesise to produce food, but that they also take in nutrients from the soil (through observing use of fertilisers).

Students can have a range of misconceptions about water and transport in plants. Some may think that water:

- enters a plant through the leaves (when students water plants, they often pour the water on the leaves)
- leaves a plant through the flowers
- taken in through the roots is retained (i.e. none is lost through the leaves)
- exits the leaves as a liquid (some may have seen guttation)
- is pumped around the plant in the same way as a heart pumps blood.

A teaching sequence

Less advanced students will be able to learn about transport in plants with only a cursory understanding of osmosis. However, to really understand transport, an understanding of osmosis is essential early on. Following that, it is sensible to follow a similar pattern to that for humans, and look at the vessels through which transport occurs. Finally, in the absence of a pump like the heart, it is important for students to understand how transport happens in plants, both of water from roots to leaves, and of sugars from the leaves to the rest of the plant.

■ Diffusion and osmosis

Osmosis refers to the diffusion of water across a partially permeable membrane. Students find osmosis hard to understand because we don't usually refer to diffusion of the solvent, and different books define it in different ways. It is important to build their understanding of osmosis gradually, so first remind them about diffusion from earlier in this topic. Then help them to understand how adding a solute can affect the concentration of water:

PROCEDURE

1 Present two large measuring cylinders of the same volume (2l or 5l cylinders). Label them A and B. Fill both with warm water to about two-thirds full (making sure the levels are identical in the two cylinders). Ask the class what will happen to the level in A if some sugar is poured in (likely answer: 'It will go up.').

2 Add 150 cm^3 of sugar to A. Ask the class what actually happened (likely answer: 'The level did go up.'). Ask the class what will happen to the level if the sugar dissolves (likely answer: either 'It will stay the same.' or 'It will go down.'). Work on this difference of opinion in discussion.

3 Shake the cylinder, or use a magnetic stirrer, until the sugar dissolves (the level does not go down). This means that the level of solution in A is now higher than the level of the water in B.

4 Pour off the extra volume from A into a small beaker until the levels in A and B are identical. Ask the class which cylinder has more sugar in it (Likely answer: A). Ask the class which cylinder has more water in it (Likely Answer: B, although some may say they're the same). Challenge those who get it wrong by pointing out the (sugary) water in the small beaker.

To complete the demonstration, point out that water can diffuse from one solution to another from a high water concentration (B) to a low water concentration (A). Having reached this point, define osmosis as the diffusion of water from high to low concentration through a partially permeable membrane. Point out that water moves in and out of cells by osmosis because the cell membrane is partially permeable.

You can demonstrate this movement using some or all of these models:

- Use Visking tubing filled with black treacle and submerged in pure water so you can see the effect of water diffusing in (the Visking tubing enlarges in size and the colour of the treacle pales as it is diluted).

- Balloon in a 'paper box' model to look at effect of osmosis on turgidity and plasmolysis (Figure 3.4a). As you inflate the balloon, it pushes on the inside of the paper box, making it bulge.
- Golf balls and marbles in a tray with an artificial partially permeable membrane – cardboard with holes big enough to let the marbles through but too small for the golf balls.
- Plastic bottle model of a cell with a partially permeable membrane (Figure 3.4b).
- ICT simulation of osmosis. Try to get one with variable speed control, or one that moves slowly so students can follow individual particles.

Figure 3.4 a Balloon in a paper box model of a plant cell. Inflation of the balloon represents water entering the cell by osmosis. b Plastic bottle model of a cell with a partially permeable membrane

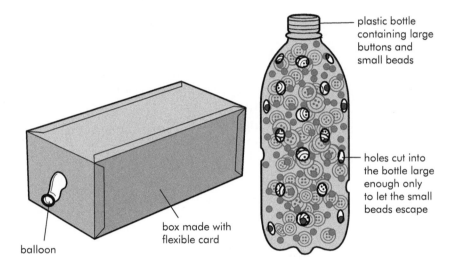

plastic bottle containing large buttons and small beads

holes cut into the bottle large enough only to let the small beads escape

box made with flexible card

balloon

Finally, show students how these models apply in living plant tissue. You could use some or all of the ideas below to help secure students' understanding:

- Measure (length or mass), bathe in water and re-measure each of ten sultanas. Because the cells in the sultanas are relatively dried out, the concentration of water in the cells is low and so water diffuses in.
- Examine slides of red onion cells or rhubarb stem cells in solutions of pure water and in concentrated sugar solution. In the former, the cell membrane pushes up against the cell wall. This is because water has diffused into the cell from a region of high water concentration (pure water) to lower water concentration (in the cytoplasm). In the latter, the concentration of water in the sugar solution is lower than the concentration in the cytoplasm. Water diffuses out of the cell and the membrane pulls away from the cell wall. Because the cytoplasm is coloured red in red onion cells, this effect is clear when viewed under a light microscope.

- Place potato chips in sugar solutions of varying concentrations; in those solutions with high water concentration (and low sugar concentration), water will diffuse into the cells and the potato chip will feel stiff. The opposite will happen in those solutions with the highest sugar concentrations (and lowest water concentration).

To assess students' understanding, you could ask them to write questions and answers about osmosis in the style of Yahoo answers (http://answers.yahoo.com/), and to construct a table that compares and contrasts diffusion and osmosis. Do be careful though; some students can end up thinking that diffusion happens in animals and osmosis happens in plants.

■ Demonstrating transport

Having secured a good understanding of osmosis, it is now sensible to demonstrate that transport does occur in plants. Many students' impression of plants is that they are relatively passive, so showing evidence of transport is important. An easy way to do this uses flowers with white petals. They will take up stains if their stalks are left in water containing coloured ink overnight. It is possible to take a thick stem of a white carnation and split it down the middle, putting each half into different coloured inks. The following day this gives a flower head that is twin-coloured and somewhat spectacular.

■ Vascular tissue

Vascular tissue comprises the vessels through which transport of water, minerals and sugars happens. A shoot taken from a plant such as Busy Lizzy or a celery petiole (stick of celery) will take up a stain (such as Indian ink) and can be used to provide material for sectioning. Students can cut thin sections of the stained stem or petiole and prepare their own temporary slides. These will demonstrate that ink travels up the stem through particular regions or tubes (called xylem). You can extend this idea to suggest to students that sugars are also transported through different tubes (called phloem).

The distribution of vascular tissue in plants is related not only to transport but also to support. In a stem the vascular bundles are arranged in a circle around the outside of the stem. In a root they are arranged at the centre of the root. A neat demonstration to explain the function of this difference involves using pieces of rolled up A4 paper to simulate the vascular bundles:

PROCEDURE

1 To simulate their arrangement in a stem, arrange the rolled-up pieces of paper in a circle and balance textbooks on them (be aware to keep students at a distance) to show that the circle can support substantial weight. (If you keep adding textbooks until the paper collapses it may be worth ensuring students move back a little!) Explain that this is one reason for their organisation in stems. (This doesn't apply to all stems, but does apply to all those you'll meet in school biology!)

2 The forces on a root are different though – this is pulled longitudinally when a plant blows in the wind. Hence, putting all the vascular tissue in the centre of the root helps resist this pull. You can show this by asking students to pull on opposite ends of a bundle of rolled pieces of A4 paper to see if they will rip, which they don't.

You may want to ask students to draw a transverse section of a stem and a root, using either prepared slides or prepared diagrams in a textbook, and to make and annotate models of the vessels using empty drinks bottles. You could ask students to research and record a video (using a digital camera or their mobile phone) about xylem vessels and phloem sieve tubes, using their models.

■ Transpiration and the transpiration stream

So what makes water move into the roots, through the roots, up the stem, and through the leaves? There is no heart, so the force that makes this happen has to come from somewhere else. Predominantly, it comes from transpiration (evaporation of water from the leaf surface). Before starting on transpiration, make sure students are familiar with changes of state (specifically evaporation).

The fact that plants lose water from their leaves can be demonstrated by placing a clear polythene bag over the shoot, but not the soil, of a potted plant for a few hours in advance of a lesson. Condensation will accumulate inside the bag.

More water is lost from the lower surface of most leaves than from the upper surface. Students can investigate this in two ways:

PROCEDURE

1 Sellotape cobalt chloride paper to the top and bottom of a leaf. The paper turns pink when damp, and because there is no wax (and more stomata – see below) on the underside of a leaf, this loses water more quickly, and the paper on it turns pink first.

2 Detach four leaves from a plant. Leave one as a control. Put Vaseline on to (i) both top and bottom surfaces, (ii) top surface only, (iii) bottom surface only of the other three. The plants without Vaseline on their bottom surface will lose water (and mass) most quickly.

You should help students relate these differences to the leaf's adaptations to prevent water loss. The wax on the upper side of the leaf reduces diffusion. This leads conveniently into an investigation of the distribution and involvement of stomata (tiny pores in the leaf) in transpiration. Microscopic examination of nail varnish replicas of upper and lower epidermal tissue, produced by the students, might be useful here. Practical details for this can be found in the websites listed at the end of the chapter (there is an investigation there which would be useful for higher attaining students). Students should be able to conclude for themselves that there are more stomata on the underside of the leaf, and synthesise the above findings to conclude that water leaves the plant through the stomata.

Given that plants wilt if they lose water, it may seem unintuitive to some students that a plant should have stomata that let water out. There are two reasons for stomata. The first is that gaseous exchange also takes place through the stomata, so they have to be open during the day to let carbon dioxide in and oxygen out. Second, letting water evaporate drives absorption and movement of water through the plant; this replaces water lost by evaporation, and enables absorption and transport of mineral ions.

If the plant does become short of water it can close its stomata. The cells on either side of the stomata are called guard cells; they are the only cells on the lower side of the leaf that can photosynthesise. During the day (when they need to open the stomata) they photosynthesise, increasing the concentration of glucose in the cytoplasm and hence decreasing the relative concentration of water in the cytoplasm. As a result, water flows into them from neighbouring cells. Most students would predict the guard cells would bulge and so close as a result of this. However, the inner wall is thick so doesn't bulge out. You can model this for students by taking two long balloons and sticking Sellotape to one side of both. Put the balloons together with the pieces of Sellotape facing each other. Inflate the balloons and they will each form a semicircle shape, leaving a hole in the middle.

The next step is for students to understand that water moves into the roots in response to loss of water from the leaf. There is an animation in the websites listed at the end of the chapter to help you explain this but, essentially, follow the path of water starting at the leaves. Explain that water is lost from the mesophyll cells, reducing the concentration of water in the cytoplasm of those cells. Water then moves into those cells from neighbouring cells by osmosis, and this repeats itself all the way back to the xylem vessels. When water leaves the xylem in the leaf, water diffuses up the xylem from the xylem vessels in the root (which you can think of as continuous with the leaf xylem). With a lower water concentration in the root xylem, water moves out of neighbouring cells, and sets up a diffusion gradient, all the way back to the root hair cells, where water moves into the root from the water in the soil. Although this explanation is simplified and the level of your explanation will depend on the students involved, you would not usually be expected to discuss this movement in terms of water potential gradients.

It is important to stress that mineral salts are also transported in the xylem, essentially swept along by the flow of water.

One way to demonstrate the transpiration stream is by using a bubble potometer, practical details for which can be found in the websites listed at the end of the chapter. Good species to use include *Buddleia*, willow (*Salix*) and willow-herb (*Epilobium*). Make sure you practise setting it up beforehand as it can be very fiddly. You can use this to investigate the effect of environmental conditions on transpiration and the transpiration stream, for example at different ambient temperatures or humidity. A simple weight potometer with computer simulation for the bubble is an alternative.

The effect of air movement can be demonstrated using an electric fan or hairdryer. Students might have the opportunity here to employ data-logging devices linked to computers to obtain continuous readings of water loss from a whole plant. Continuous readings of the mass of a plant on a balance experiencing known changes in environmental variables can produce some interesting graphs for interpretation. There is a simulation in the websites listed, which also does the same job.

The penultimate step is to look at the root hairs as being adapted to maximise diffusion of water and transport of mineral salts into the root. You could ask students to examine the root hairs produced by cress seedlings, germinated in advance of the lesson, using a hand lens or a low-power microscope. To assess their understanding of the role of root hairs, you could also ask students to write a mission statement for the root, to include an explanation of how exchange occurs. Or to synthesise their understanding of root hairs with other adaptations to increase surface area, they could write a lonely hearts column for an efficient exchange surface (e.g. efficient exchange surface wanted, gsoh ...). Finally, students could build a

social photo-sharing site (on a site like www.flickr.com), which they use to educate the reader about the role of root hairs.

The final part of the story is examining how mineral salts enter the root hairs. If they diffused in passively the plant would not get enough, so they have to be pumped in actively (using energy) through tiny pumps on the root hair membranes. To assess their understanding of the differences between active transport and passive diffusion, you could ask students to write a poem, or to design their own role play to perform to the class. Secondary data are available for analysis in the websites listed at the end of the chapter.

■ Translocation

Getting first-hand evidence of the involvement of phloem in transport is difficult at this level. However, students may be aware of the damage that occurs to trees and shrubs if their bark is 'ringed' (see Figure 3.5). They may well have seen young trees in woods or parks with protective sleeves around them to prevent their bark being damaged. You could also provide the group with a piece of continuous prose about translocation, written to suit their reading level and learning expectation, and set some specifically targeted questions to extract the desired information. At this level it is probably sufficient to bring out the following points:

- Phloem cells are alive.
- Their walls are permeable.
- Phloem is involved in the transport of sucrose (formed from photosynthetic glucose) and other organic molecules.
- Transport in phloem occurs both up and down the plant, with the leaves as the source of glucose.

Figure 3.5
Removing bark and phloem prevents downward movement of sucrose, which accumulates above the ring

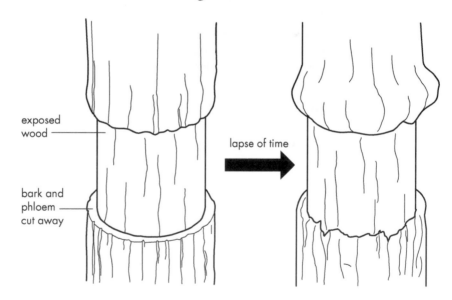

exposed
wood

lapse of time

bark and
phloem
cut away

Further activities

- Students could research the ways in which plants are adapted to desert environments, reporting their findings as a podcast 'from our own correspondent'.

Enhancement ideas

- Many of the ideas above can be set in a commercial context, with the physiology of transport related to commercial growing of crops and maximising profit.
- You could present scenarios to students, which they have to make a judgement on, based on their knowledge of transport. Examples could include a planning application to fell trees and build houses in a wetland area, or a proposal to use plants to absorb toxic minerals from the soil.

3.4 Evolution of transport systems

Previous knowledge and experience

Remind students about the microorganisms you looked at earlier on. Students will understand that the organisms they had been looking at are very small, and that the maximum distance over which exchanges take place is extremely small. They should realise that these organisms do not have specialised transport systems, although they may not immediately understand why.

A teaching sequence

■ Surface area to volume ratio

The concept of decreasing relative surface area with increasing volume is always tricky for students. Plasticine models, which can be rolled out into different shapes, are a nice way of getting over the problem. The changes that take place in the surface area to volume ratio, as organisms of increasing mass and volume are considered, can also be demonstrated quickly using model animals represented by cubes – where doubling the linear dimension increases the surface area by a factor of four but the volume by eight. The area of surface available for exchange and diffusion alone is enough to supply the needs of a unicellular organism. The increasing volume and distance from deeply seated cells to the relatively smaller exchange surface make diffusion alone too slow and haphazard as the sole means of exchange for larger organisms, especially as some larger organisms are very active. The evolution of transport systems that give faster delivery, and delivery with direction, has overcome both these inadequacies.

Students might carry out an investigation into the effect of decreasing surface area to volume ratios on surface exchange using agar or gelatine cubes and an aqueous solution of a food dye. Cubes of differing sizes immersed in the dye could be sectioned after a fixed time to see the extent of 'delivery' by diffusion alone. You will need to have tested this earlier to find out when to section the cubes to give a clear difference. Full details of a protocol are available in the websites listed at the end of the chapter.

Further activities

- Students might be encouraged to think about other problems for living organisms that may be associated with an increase in size. Examples could include a review of the way gases are exchanged in larger organisms, how temperature control is achieved and the way physical support is maintained.
- Students could do some individual research on the size of different living organisms. They could find out which are the biggest organisms, plants as well as animals, in different types of habitat. If they have access to appropriate data they might compare the maximum size shown by terrestrial and aquatic organisms from the same phylum.

Other resources

Background reading

Griffths, H. and Howard, A. (1997). Roger the red blood cell. *School Science Review*, **79** (286), pp. 101–103.

Lester, A. and Lock, R. (1998). Sponges as visual aids – bath time fun for biologists. *Journal of Biological Education*, **32**, pp. 87–89. This article describes teaching aids using synthetic sponges as models of red blood cell and the heart.

Websites

Experiments on diffusion: www.practicalbiology.org/areas/intermediate/exchange-of-materials/diffusion/

Animation of diffusion: www.johnkyrk.com/diffusion.html

Protocol for dissection of a heart: www.practicalbiology.org/areas/intermediate/cells-to-systems/structure-of-a-heart/looking-at-a-heart,76,EXP.html

Animations of red blood cells: www.youtube.com (search for 'red blood cells')

Safety advice about using blood: www.cleapss.com

Virtual heart dissection: www.gwc.maricopa.edu/class/bio202/cyberheart/anthrt.htm

Practical experiments to investigate osmosis: www.practicalbiology.org/areas/advanced/exchange-of-materials/osmosis/

Making a leaf peel of stomata: www.practicalbiology.org/areas/advanced/exchange-of-materials/transpiration-in-plants/measuring-stomatal-density,71,EXP.html

Animation about transpiration: http://croptechnology.unl.edu/animationOut.cgi?anim_name=transpiration.swf

Measuring rate of water uptake using a potometer: www.practicalbiology.org/areas/advanced/exchange-of-materials/transpiration-in-plants/measuring-rate-of-water-uptake-by-a-plant-shoot-using-a-potometer,62,EXP.html

Simulation potometer to investigate effect of environmental conditions on transpiration: www.sycd.co.uk/can_we_should_we/explore/plants/potometer-applet.html

Animation to show the transpiration stream (more detail than required, but good for making the journey of water clear): www.biologymad.com/resources/transpiration.swf

Secondary data for analysis about active uptake of mineral ions: www.practicalbiology.org/areas/intermediate/exchange-of-materials/active-uptake/

Effect of size on uptake by diffusion: www.practicalbiology.org/areas/intermediate/exchange-of-materials/diffusion/

4 Gas exchange, movement and fitness

Ann Fullick

4.1 Gas exchange
- Knowledge and understanding of the thorax and breathing system
- How breathing and gas exchange take place, relating structure to function
- Effects of tobacco smoke/pollution/allergens on the breathing system

4.2 Respiration
- Need of all living cells for usable energy source
- Aerobic respiration
- Anaerobic respiration

4.3 Skeleton and muscles
- The roles of the skeleton and associated tissues – bone, cartilage, tendons, ligaments and muscle, in joints and in movement

4.4 Exercise and fitness
- Fitness as an example of the way different body systems work together
- Anaerobic exercise and oxygen debt
- The changes in heart, lungs, skeleton and muscles in response to increased exercise

Choosing a route

These topics can be taught as part of a linked and integrated whole. This can work very well, giving a good example of integration of organ systems and purpose behind learning – doing a practical on exercise that makes a student puff makes it very clear why breathing matters and also what fitness is all about! On the other hand, the topics can be taught in smaller linked groups – you might choose to link gas exchange and respiration; or skeletons and fitness; respiration and fitness; gas exchange, respiration and fitness – or as independent topics. The route you choose will depend on a variety of things including:

- the demands of the curriculum
- the demands of the specification followed with older students
- the order in which other topics are to be introduced
- personal preference
- departmental policy.

Any combination will be effective if presented in a lively and enthusiastic way!

Students often get confused between breathing, gas exchange and respiration. Whilst teaching these topics it is helpful to be constantly aware of this and to try and keep the different concepts separate in their minds. Some teachers prefer to do this by dealing with breathing and gas exchange at a different time to respiration. Others prefer to help students recognise with great clarity the way the different processes are dependant on each other. If a student really understands that breathing movements get air into the lungs so that gas exchange can take place, and that the oxygen brought into the body in this way is needed in the cells for respiration, then the likelihood of confusion is much reduced. Examination specifications often place breathing, gas exchange and respiration in the same section, so it may be that you deal with the topics separately for younger students and together when teaching exam groups.

One of the joys of teaching these parts of the curriculum is that students can use their own bodies as an experimental system. As long as teachers are aware of safety concerns and of the need to control any unhelpful and competitive comparisons of performance, this is an area of biology which can be full of interest for students because it is so easily related to themselves, their families, their peers and sporting personalities.

 Be aware of medical conditions that affect the ability of certain students to perform exercise, e.g. students with asthma may need to use their inhalers first. Exclusions from practical work need to be handled carefully to avoid 'labelling' students with disabilities. However, all students breathe, and most students can increase the rate at which they breathe with some sort of exertion, so the great majority of students can be included in some way.

One final point is that as teachers we need to keep remembering the demands of a spiral curriculum. Students are often revisiting a topic when we introduce it to them, or we in turn may be revisiting it later in their school careers. Thus, when planning these topics for younger students it can be helpful if we also think about what they will be taught later. Tempting as it can be to tell all our best anecdotes and introduce all the really fascinating ideas to our students lower down the school, if we ration them and keep some things fresh for later we – and our students – will reap the benefit.

4.1 Gas exchange

Previous knowledge and experience

Students will probably have met relatively little material specifically linked to breathing and gas exchange in primary schools. They are likely to have met the lungs as one of the organs of the body, and will know that living things need oxygen and produce carbon dioxide as a waste product. A common misconception at this level is that we breathe in oxygen and breathe out carbon dioxide! This can be cleared up quite easily by asking students what makes up the air. When that has been established you can demonstrate that air is what we breathe in and breathe out – the balance of the gases in it has just been changed slightly.

A teaching sequence

Before beginning to teach this topic you need to decide your desired learning outcomes – in other words, where you hope your students will be by the end of the topic. These will tend to vary depending on the age and ability of the students you are working with, but if you have a list to start off with it gives you a tool for evaluating the success of what you do. For example, with younger students you might be quite happy if they know and can label the main parts of the breathing system, but with older students you might be hoping that you will also have given them an insight into the physical effects of exercise, smoking and altitude on that system, and why the body reacts as it does. It is always worth having high expectations of what your students will be able to take away from this topic. It has lots of intrinsic interest, and if you expect a lot from them you may be surprised at what they can achieve.

It is a good idea to start off with some questions, which will give you a feel for what the students have met before: 'Do they know that people need to get oxygen into their bodies and get rid of waste carbon dioxide?' 'Do they know that other animals and – very importantly – plants have the same requirements?' 'Do they have a feel for the idea of the gases being exchanged in a specific organ?' 'Where do they think breathing comes into all of this?' etc. By planning your questions carefully you can find out quite a bit, and if you can structure things so you get a feel for the knowledge of the whole class rather than one or two individuals, so much the better

('Hands up everyone who thinks plants don't need to bother with getting oxygen' 'Do plants have to get rid of carbon dioxide?' 'Who thinks breathing helps us get rid of carbon dioxide?'). It is often particularly useful to go into the first lesson with plenty of material prepared. If your feeder schools or colleagues lower down the school have covered this topic in great detail and the students have a really good grasp of the basics you were planning to deal with, it is always helpful if you know that you have more than enough prepared to keep them occupied and interested. If the first lesson of the topic interests the students, they are much more likely to view the coming lessons in a positive light. It can be useful in the first session to give them an overview of what they will be looking at in the topic, and why it is exciting and relevant to them. This can help students to see a purpose to what they are learning and give them a sense of how the parts of the topic interlink and support each other.

■ Social, moral and ethical issues

Wherever possible, everyday analogies can help students grasp ideas – you'll find some examples in the different areas of the topic covered below. Students often love the stimulus and challenge of social, moral and ethical issues associated with a syllabus area; they will wrestle to understand the science to help them argue a case for or against an idea, and if you read a daily paper or a magazine it's useful to cut out articles linked to topic areas like this when you see them. For example, material on asthma, cystic fibrosis, health risks from air pollution and smoking often appear in the press. You can then use these in teaching as a basis for a discussion, a class exercise, a homework exercise or a starting point for individual research.

■ Physical principles

There are four physical principles that are needed for complete understanding of gas exchange:

- diffusion
- surface area
- surface area : volume ratios
- air movement in response to pressure changes.

It is worth checking with colleagues to find out if you can reasonably expect some understanding of these areas from work done in other subjects such as maths or other science topics. However, it is a good idea to run a quick experiment or demonstration of each when you reach the appropriate place in the teaching sequence to make sure students really do have a clear picture of what is going on.

There are lots of easy demonstrations of diffusion, from potassium manganate(VII) crystals dropped into water to a perfumed spray squirted in one corner of the laboratory, see Chapter 3. It is useful to emphasise that diffusion takes place down a concentration gradient from where there are a lot of particles of the diffusing substance to where there are relatively few. This is also a useful opportunity to point out that diffusion 'just happens' – that it is due to the random motion of particles and is not an active, energy-consuming process.

The following exercise is useful to help students grasp the idea of the alveoli providing the lungs with a large surface area for gas exchange. It also helps them understand the problems that result in chronic obstructive pulmonary disease (COPD, which used to be known as emphysema) when the alveoli break down to form big air sacs:

1 Provide groups with one large potato and several little ones with (between them) approximately the same mass as the large one.

2 Ask the students to peel the potatoes and use graph paper to work out the approximate surface area of the peel of each mass of potatoes. The small ones should (between them) demonstrate a larger surface area – like alveoli.

3 Students must be reminded not to eat any of the raw potato because of the risk of contamination in the laboratory.

Surface area : volume ratio can be demonstrated with cubes (either drawn on the board or real) with sides of different lengths.

For demonstrating the movement of air in response to pressure changes, either use a bicycle pump or a model chest (many school laboratories have a 'balloon in a bell jar' model chest readymade for this part of the course (Figure 4.1). It can be useful to have a single balloon and present this simply as a model of the way air moves in and out of the alveoli as the pressure changes, rather than to have two balloons set up as model lungs. This is because many students already have the misconception that their lungs are like a pair of balloons inflating and deflating in their chest, and the traditional 'model chest' reinforces that misconception. By having a single balloon it can be presented as a single air sac, one of millions within your lungs, whereas students have a marked tendency to interpret two balloons as two lungs. A cut open pink bathroom sponge is useful too; show them the spongy structure of air sacs and then explain that the balloon represents one sac.

Figure 4.1 An artificial chest with a diaphragm in different positions and air moving into or out of the balloon. (Try the apparatus out beforehand. All too often the seals are leaky, the balloon has perished and it will not work! Well put together, it can be good.)

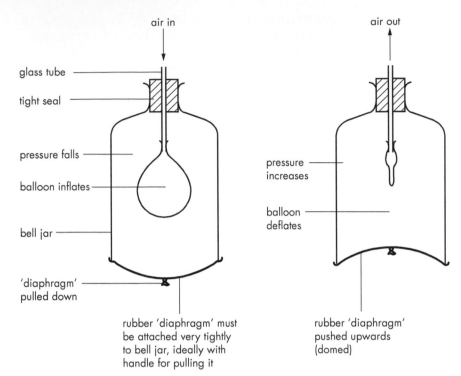

air in

glass tube

tight seal

pressure falls

balloon inflates

bell jar

'diaphragm' pulled down

rubber 'diaphragm' must be attached very tightly to bell jar, ideally with handle for pulling it

air out

pressure increases

balloon deflates

rubber 'diaphragm' pushed upwards (domed)

■ The need for gas exchange

Living things need energy to carry out all the reactions of life. To release the energy from their food and transfer it to a usable form requires oxygen in respiration. During the release of energy from food, carbon dioxide is produced as a waste product. It builds up in the cells and must be removed from them because it is poisonous. Living organisms need oxygen and need to get rid of carbon dioxide, so they need to exchange the gases. Tell students that this is the same principle as changing an empty gas cylinder for a full one when they're camping, or returning a book they have read to the library and exchanging it for a new one they want to read. It's useful and important to emphasise here that gas exchange takes place in living organisms, whether or not they breathe. This is one point at which you might choose to bring in the idea that gas exchange takes place as a result of diffusion, and clarify diffusion as described above.

One common misconception is that pure oxygen is moved into the lungs and exchanged for carbon dioxide, which is then breathed out. Always reinforce the idea that air moves in and out of the lungs, and the oxygen and carbon dioxide that are exchanged only make up part of the air. You can give students a table showing them the difference in oxygen and carbon dioxide concentration in inhaled and exhaled air to make this point more clearly.

■ Organs for gas exchange

The idea that some organisms need specialised organs to bring about successful gas exchange is an important one. Plants have low energy requirements, and make their own oxygen in photosynthesis, so they don't require special ventilated organs. Very small animals can obtain sufficient oxygen to supply their needs by simple diffusion. But as animals get bigger the diffusion pathways get too long to supply all the oxygen they need to support their much more active lifestyles. The concept of surface area : volume ratio (see page 91) and its implications for breathing systems as animals get bigger and more active is an important one to get over.

■ Gas exchange in humans

The important aspects of the process of gas exchange in humans are:

1: Structure of breathing system

A model chest with removable parts can be used to make it easier to understand how the breathing system and thorax fit together. Get across the spongy structure of the lung tissue. Particularly with older students, it might be worth either letting them use microscopes themselves to look at slides of lung tissue, having slides projected on to screens or using an appropriate CD-ROM or web resource to make this clear. The cartilage rings of the trachea and the way food is swallowed past them often causes interest, and the protection mechanism of the body to prevent food going down the windpipe into the lungs is one most people have experienced, even if they have not understood what is going on. A brief explanation of why food in the lungs is damaging, and why choking can cause death, can lead to a brief but valuable description of how to help someone who is seriously choking. This is an aside that can save lives!

With older students, the beating of the cilia of the ciliated epithelium in the trachea, etc. can usefully be mentioned here, not least because they will be referred to later when talking about smoking.

2: Method of ventilation

This involves looking at the principle of air moving in response to changes in external pressure. There are a number of ways of approaching this.

If you start off by simply asking the students how they breathe in and out, you will know what their misconceptions are. You can then describe what happens in the body when we breathe in and out. Putting their hands on their own ribcages and taking a deep breath in and out usually enables students to work out how the ribs move. Envisaging the intercostal muscles, which move the ribs, is helped if they think of barbecued spare ribs or rib of beef and realise that what is eaten are the muscles that move the ribs (although these examples might not be appropriate when in a multicultural environment, where, for example, students who are Hindus, Buddhists or Jains would not eat these foods for religious reasons).

The movement of the diaphragm as it flattens will usually need describing. The effect these movements have on the volume of the chest and so on the air pressure inside the chest can then be explained. Then, using the demonstration in Figure 4.1, the passive movement of air into and out off the lungs can be discussed. Students should recognise that in normal quiet breathing, breathing in is an active process (energy is expended on muscular contraction) whereas breathing out is passive as the intercostal and diaphragm muscles simply relax, reducing the volume of the chest cavity and so increasing the pressure of air in the lungs and forcing it out of the system. They will need to know that breathing out can be active too.

3: Exchange process in the alveoli/adaptations to function

This is where the concept of diffusion becomes very important (see page 91). Students need to know the structure of an individual alveolus and its close association with blood vessels. If they have not yet looked at the circulatory system, it is a good idea to flag here the fact that having got oxygen into the body, it is important to be able to carry it to the cells where it is needed – hence the blood supply as a transport system. It is difficult to talk about the structure of the alveolus and its capillaries without dealing with the issues of adaptation for function. Looking at the shape of the alveoli, single cell layers and the close proximity of the blood vessels makes it easy to point out how this system is so well adapted to the movement of oxygen out of the air in the lungs into the blood and the movement of waste carbon dioxide from the blood into the lungs:

- large surface area for gas exchange to take place (see Physical principles)
- thin alveolar walls (single cell thick) so short diffusion pathway
- rich blood supply to carry carbon dioxide to the lungs and oxygen away from the lungs, maintaining a concentration gradient to aid diffusion in both directions.

The concentration gradient, which is maintained by blood flowing through the vessels and the changing of the air in the lungs, can be explained to older and more able students.

There are a large number of video clips and animations available, which demonstrate both the ventilation movements of the chest moving air into and out of the lungs and the process of gas exchange in the alveoli. However, many reinforce common misconceptions by talking about oxygen coming into the lungs rather than air which is rich in oxygen, and carbon dioxide leaving rather than air with a higher concentration of carbon dioxide. They may also show the lungs inflating and deflating without showing the movements of the ribs and diaphragm that bring this about, which can lead students to think that the lungs self-inflate. So a useful exercise is to ask students to find two or three of these online resources and look at them with a critical eye, writing a review to find the best resource to recommend as a teaching aid.

You can let students observe the effect of gas exchange in their own breathing using the apparatus shown in Figure 4.2. Either lime water (a clear liquid that turns cloudy when carbon dioxide reacts with it) or hydrogencarbonate indicator solution (a red liquid that turns yellow when carbon dioxide dissolves in it) can be used as indicators. (As long as the experiment is not continued for too long, the lime water in the exhaled tube goes cloudy and the lime water in the tube they have inhaled through does not. Stop once this has happened, because eventually the carbon dioxide in the inhaled air makes the lime water go cloudy and too much carbon dioxide causes the clouding in the exhaled tube to disappear.)

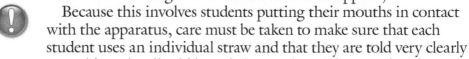

Because this involves students putting their mouths in contact with the apparatus, care must be taken to make sure that each student uses an individual straw and that they are told very clearly to avoid getting liquid into their mouths. Make sure that the glassware is clean.

Eye protection is needed when handling lime water.

Some schools suggest doing this with the apparatus joined together so students breathe in and out, squeezing various bits of tubing to direct the flow of air. The apparatus shown here is simpler, avoids confusion and also helps to prevent much spluttering and lime water getting everywhere!

Figure 4.2
Apparatus to show
testing of inhaled
and exhaled air for
carbon dioxide

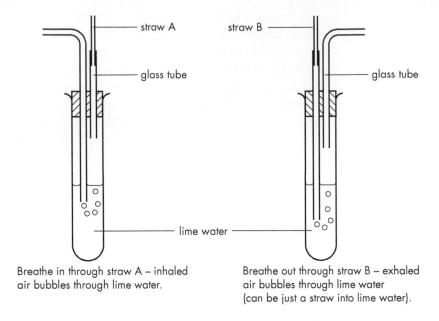

Breathe in through straw A – inhaled
air bubbles through lime water.

Breathe out through straw B – exhaled
air bubbles through lime water
(can be just a straw into lime water).

■ Effect of smoking/pollutants/allergens on the functioning of the breathing system

The effects of smoking on the lungs and the way they work can be
dealt with very effectively here as part of the work on gas exchange,
or they may be covered when dealing with the effects of drugs on
the body as a separate topic. You may also be asked to support work
on this in PHSE or similar courses. One advantage of looking at
the effects of smoking at this point is that it can be tied in with the
effects of other air pollutants and of allergens in causing asthma.

The main effects of smoking on the breathing system include the
following:

- Smoke anaesthetises the cilia in the trachea and bronchioles,
  allowing mucus, bacteria and dirt to accumulate in the lungs and
  so making smokers more open to infection.
- Smoke contains a number of known carcinogens (chemicals that
  can cause cancer/increase the risk of developing cancer), which
  can trigger changes in the cells of the lungs, turning them
  cancerous. Because the lungs are a large space with few sensory
  nerve endings, cancers can grow until they almost fill one of the
  lungs and have spread widely around the body before they make
  the individual so ill that they go to a doctor and are diagnosed.
  Because of this, lung cancer can often be fatal.
- Tar and other chemicals that are part of cigarette smoke build up
  in the lungs on the surface of the alveoli and make gas exchange
  less effective.

- Many alveoli break down in response to the irritant chemicals in smoke, leading to fewer, bigger air spaces. There is less surface area for gas exchange to take place and the large spaces may fill with fluid – a condition known as chronic obstructive pulmonary disease (COPD).
- Cancer of the mouth, throat and lungs is much more common in smokers.

The smoke from a cigarette can be drawn through a simple filter of glass wool – students are often appalled at the level of muck collected from even mild cigarettes. Note: this experiment is not allowed in Scotland. In other parts of the UK it must only be done in a fume cupboard.

Care is needed when handling glass wool.

Many of the problems with the breathing system caused by tobacco smoke are also seen as a response to air pollution of various other types. Breathing in dust from industrial processes used to be a common cause of lung cancer and COPD in this country, and is still a risk in some parts of the world. Health and safety legislation in the workplace has made this much less of an issue in developed countries.

Industrial pollution and more everyday allergens such as grass pollen, the faeces of house dust mites, and pet hairs can affect the breathing system in more immediate ways. In sensitive individuals they trigger a release of histamine from the cells lining the respiratory tract, causing the tissue to swell and narrowing the tubes leading down to and into the lungs. This in turn increases the resistance to air flow, making it very difficult to move air into and out of the lungs and giving the symptoms of asthma. This can be treated in a number of ways, most commonly by inhalers, which deliver a small dose of adrenergic drugs to the lining cells, causing an almost immediate dilation of the tubes and easing of the breathing. Asthma may not be on the curriculum or specification, but studying this very common health problem can be a valuable life lesson for students and also reinforce learning of the required content. Some sports people are successful even though they suffer from asthma because they know how to relieve it.

■ Treat your students sensitively

When dealing with health aspects like these it is important to be very sensitive and non-condemnatory. Many students will have parents who smoke, and they may feel great concern for their health. When dealing with asthma, a number of students in most classes will carry asthma inhalers and will suffer from the problem to a greater or lesser degree. If the students are confident and outgoing, and you have a good relationship with the class, some of your students with asthma may be prepared to explain to their classmates what an asthma attack feels like and how they use their inhaler. However, it is never a good idea to spring a request to share personal information on a student in the middle of a lesson. The end of the preceding lesson is an ideal time to have a quiet word, explaining that you will be covering asthma in the next lesson and asking if the student/s are happy to make some input.

■ Data research

This part of the course offers many opportunities for students to research and discuss the implications of the scientific knowledge. There are several possibilities for looking at aspects of How Science Works (HSW). The internet, libraries and doctors' surgeries are all places where students can find out for themselves about the effects of smoking, asthma and air pollution. There are a lot of data available on the internet on the mortality and morbidity (ill health) of smokers, which you could present to students for analysis and comment. Questions such as: 'Why are air quality and the pollen count now regular features of weather forecasts in the summer months?' 'Why do people think that the number of people with asthma has gone up and up at a time when most diseases have been getting less common?' can be posed for research, thought or discussion.

Articles or data on smoking from the media can be looked at, and reasons why young – and older – people continue to smoke in spite of widespread awareness of the health risks can be discussed. Issues such as the damage to the health of non-smokers by passive smoking (breathing in smoke-filled air from smokers) can be raised, as well as the rights of smokers to smoke against the rights of non-smokers to clean air. The impact of smoking bans in many workplaces, as well as in restaurants and pubs, on public health could be discussed in terms of scientific evidence for the benefits, individual freedoms, etc.

Another question that can be raised is the increase in the number of people, and children in particular, who are diagnosed with asthma. Are people sometimes diagnosed as asthmatic and given inhalers now when in the past they would have managed without them, simply because the drugs are there and drug companies encourage doctors to prescribe them?

This area of biology really is an excellent opportunity to set up discussion, role plays and other forms of interactive involvement for the students. As long as the activities are focused and there is an element of reporting back findings or ideas within a set time, this is often a remarkably fruitful experience for students across a wide range of ages and abilities. It can be used to reinforce the HSW skills of questioning and debating the evidence used in decision-making and the way scientific evidence influences society. It also provides an opportunity to look at the difference between causal links and associations by drawing on the data on smoking and cancer.

A common misconception amongst students is that exercise/sport will undo the effects of smoking. This is not true. Exercise will make the very immediate effects of breathlessness less noticeable by increasing the size of the heart and lungs. However, smoking means that the benefits from this will not be as great as they would otherwise have been, so a person's sporting performance will not be at its maximum, and the problems of addiction, smoke damage to the lungs and carcinogens are exactly the same whether the person exercises or not.

■ Gas exchange in other living organisms

Gas exchange is vital to living organisms. It is important that students recognise how and where it takes place in plants (particularly the spongy mesophyll) and how the use of carbon dioxide and release of oxygen by plant cells as a result of photosynthesis reduces the need for specific gas-exchange systems. For completeness and interest a lightning tour of some of the ways in which a variety of animals other than mammals manage gas exchange can also be interesting and fun. Fish (gills), insects (tracheae), frogs (skin, simple lungs), etc. are obvious examples, and more exotic creatures, such as those insect larvae that breathe through their bottoms, are always popular! A useful exercise is to look at each of the gas-exchange systems with reference to the way they are adapted for their function, so that as well as broadening the students' knowledge of a function common to all living things it reinforces the 'large surface area, thin walls, concentration gradient' ideas that are so fundamental to an understanding of gas exchange.

Further activities

- Response of the breathing system to exercise and increased oxygen demand.
- Adaptations of the gas exchange system in response to regular exercise, living at altitude, etc.
- You can use sensors and data-loggers to help measure the effect of exercise on the breathing rate, the strength of the lungs, the levels of oxygen and water in inhaled and exhaled air, etc.

These topics can be brought in and dealt with here, or in a separate section looking at the concept of fitness, which might lead on from this work on gas exchange, or might be used to link and reinforce work both on gas exchange and respiration. The response to exercise can also be treated independently under a more general 'Keeping healthy' umbrella, which might also include diet, infectious diseases, etc.

Enhancement ideas

- Work can be done on diseases of the breathing system, looking at ways in which diseases are spread and the role of social improvement, drugs and vaccination in the reduction of many such diseases.
- There are great opportunities for HSW, both for data handling and for developing an awareness of how selective use of data can slant a picture (e.g. figures for the incidence of TB in the UK, where the decline can appear to be almost entirely due to the introduction of a vaccination programme and antibiotic treatment if only the latter part of the twentieth century is considered, but which was in fact already occurring as living and working conditions improved over the whole of the century).

4.2 Respiration

Previous knowledge and experience

Different people will chose to teach respiration in different places in their courses. In this book it is dealt with in detail in Chapter 1 as one of the fundamental processes taking place in cells. If, in your school, respiration has been dealt with before you tackle it here, a quick résumé, as suggested below, is very useful to support the next section of work. This looks at the demands of the body during exercise, when levels of gas exchange and breathing change to supply the changing demands of cellular respiration. However, if respiration has been dealt with only in passing earlier in the course, you may choose this point to cover the material fully. In any event, reinforce the linked nature of processes within humans and other organisms.

A teaching sequence

Crucial ideas about respiration, which the students need to have grasped to make sense of the work on exercise and fitness are:

- Food taken into the body contains stored chemical energy, which needs to be transferred to a form that can be used by the cells of the body during respiration. This aerobic respiration needs oxygen, which is supplied by breathing and gas exchange.

- glucose + oxygen → carbon dioxide + water + *usable energy*
- At rest and during gentle activity, normal breathing supplies enough oxygen for respiration to take place.

Ideas on respiration, which you might wish to revisit briefly with more able students before embarking on work on fitness include the following:

- Respiration occurs all the time in living organisms.
- Aerobic respiration takes place in special cell organelles called mitochondria. They contain all the enzymes needed to produce useable energy from glucose using oxygen.
- If there is insufficient oxygen reaching the cells, then the enzymes of the mitochondria cannot function properly and *anaerobic* respiration takes over.

4.3 Skeleton and muscles

Previous knowledge and experience

Students will have met the idea that humans need food to survive and move around and that they have skeletons and muscles to support their bodies and to help them move. They may well have done projects on skeletons and made moving models of skeletons, etc. depending on the enthusiasm of their primary teachers. Be prepared to value and use readymade 'experts'!

A teaching sequence

Before dealing with issues of exercise and health it is important for students to have an understanding of how the body moves. The skeleton and how it is moved by the muscles makes an interesting topic in its own right, and if this work is then used immediately along with knowledge from their work on gas exchange and respiration to look at anaerobic respiration, exercise and fitness, a well-rounded and integrated picture of real bodies working, rather than just theoretical systems, can be developed. The skeleton can also be used as an example of an organ system in the body, with each bone containing different tissues working together to form an organ.

Even if the skeleton and muscles are not covered in great detail, students will need some introduction to the ideas to understand such topics as arthritis, ageing and wear and tear, the way muscles are affected by exercise, the way reflexes bring about change in the body, etc.

■ Background knowledge

As before, at the beginning of each unit of teaching you need to develop your own desired learning outcomes for the students. Give a broad overview of the work to come and ask questions to find out the level of prior knowledge of the group you are teaching. This may be in the form of a quiz, or a questionnaire, or simple questions as you talk; this will set the scene for the topic to come and enable you to pitch in at the right level. Again, this topic has lots of opportunity for individuals to use themselves as experimental organisms, and has considerable intrinsic interest.

■ Structure and function of the skeleton

There are a number of ways to introduce students to the structure and function of the skeleton. One is to imagine people without a skeleton – as large pink blobs – and then work out what a skeleton makes possible. Perhaps the most direct – and certainly very effective – way is to go straight in with a full-sized human skeleton (real or plastic). Put yourself and the skeleton at the front of a semicircle of students and talk through the different bits of the skeleton and what they do. The more this is sprinkled with anecdotes and explanations, the more effective it will be. Don't forget to ask students to let you know if they have broken any of the bones you are looking at and to pause for questions at regular intervals. Students are often quite fascinated by this part of the course and have many interesting observations and contributions to make, so it is worth keeping an eye on the clock so that you manage to cover the whole skeleton in the time available to you. Some of the information snippets which help things to stick in students' memories include:

- the idea of the skull protecting the brain, but with joints that allow the head to be squashed during birth, so the baby can be squeezed through the pelvis and allow the brain to grow after birth
- the fact that open joints between the bones of the skull, which allow the baby to be born, only grow together and close up slowly as the child grows to allow the brain to grow and get bigger; the soft spot (fontanelle) of babies; ageing ancient skeletons by looking at the skull joints, etc.
- orbits – deep and strong to protect the eyes
- no nose bone – nose made of cartilage
- only the lower jaw moves
- the different type of movement at the shoulder and the elbow
- ribs – stick your hand on to or – even better – into the ribcage to show how effectively the lungs, heart, etc. are protected; remind

students of the position of the intercostal muscles and the diaphragm and how these are used for breathing movements

- vertebrae – where the nerves come out, trapped nerves, cartilage discs and how they become compressed during the day so people 'shrink' by about 1 cm between the morning and the evening, and permanently compressed with increasing age
- why a broken neck or spine is so dangerous
- pelvis – the way the bones are fused; the coccyx as the tail remnant with all the nerves intact, which is why it hurts so much to fall on it; sexing skeletons by the shape of the pelvis; the gap through which a baby is delivered and why it is important that all the ligaments holding the pelvis tightly together relax in late pregnancy; protective role of pelvis for reproductive organs, etc.
- hip joints – where replacement surgery takes place
- femur – size and weight; how the length of a long bone from an ancient skeleton can be used to determine the height of the person
- ankles and feet/wrists and hands – the number of small bones needed for them to function properly.

There are many other special aspects of the skeleton, of course, and students generally find it very rewarding to be told these 'added extras'. Throughout, you can usefully and repeatedly point out examples of the main roles of the skeleton: support, protection and movement. It is usually more effective in both time and effort to provide students with ready drawn skeletons to label rather than getting them to draw their own. However, they may appreciate the opportunity to handle a particular bone – skull, vertebra, long bone or pelvis – look for special features on it and then draw and label it. This type of activity often works best if a time limit is set.

■ Physical principles

When dealing with skeletons and movement, the most important physical principle for students to get to grips with is the payoff for living things between load-bearing capacity, strength of materials and mass. Students can make models to show that hollow cylinders support almost as much weight as solid ones but are much less massive, and then look at sliced-through long bones to see how this relates to bone structure. Big bones need very big muscles to move them – this can lead to interesting discussions about what actually limits the size of terrestrial animals.

The bone is an important living tissue. It is a common misconception amongst students that bone is the sort of dead, white material we see in a preserved skeleton. It can be quite thought provoking for them to realise that their skeletons are structures that are constantly being built up and broken down. Whenever you reduce exercise levels, bone density is lost – bone cells are removed – and as soon as you carry out new or different types of movement, extra bone is laid down to strengthen the existing structure. Most bone growth takes place in children and adolescents. In young people the long bones have an area of cartilage at each end (known as the epiphyses) and this is where growth in length of the bones takes place.

After the adolescent growth spurt these soft regions become hard, calcium-containing bone like the rest of the skeleton and from that point on no further growth in height is possible. If there is a period of starvation or serious illness during childhood, bone growth is halted until conditions improve, and this halt in the growth remains marked forever as a pale horizontal line (a Harris line) in the structure of the long bones. This is used in archaeological analysis to give information about diet and disease in people who died hundreds of years ago.

It is also useful to mention osteoporosis – perhaps bringing in leaflets on the problem from your local doctors' surgery. Students need to recognise that this condition is not just a problem of old age, it affects men and women (although women are more likely to suffer from it) and that plenty of calcium and vitamin D in the diet as well as regular load-bearing exercise can help to prevent the problem arising. This can be used to revise/tie into work on diet and health carried out earlier or coming later in your course.

■ Structure and function of joints

The skeleton could perform its functions of protection and support without allowing any movement to take place at all. The joints are there so we can move, and different types of joints allow different levels of movement. The groundwork for this work can easily be done when looking at the skeleton, and the skeleton can be used again to remind students of the different types of movement the different joints make possible.

Many specifications look at the link between exercise and health. If students understand the importance of cartilage and synovial fluid in ball-and-socket and hinge joints they can make far more sense of sporting injuries and arthritis. They can understand and recognise problems with the joints and the consequences this has. As sporting injuries often affect students or athletes they know of, they are a particularly useful learning tool and help students understand the importance of the tissues in healthy joints.

■ Muscles and movement

The texture and appearance of muscle can be demonstrated with a piece of red meat – students can tease it out and look at it under a low-power microscope. Using chicken limbs, you can demonstrate clearly the way muscles are attached across joints and to bones by tendons.

In both cases hygiene must be rigorously observed, and if you do demonstrate with raw chicken it is important that your hands, laboratory surfaces and all the equipment used must be scrupulously cleaned afterwards to avoid contamination by *Salmonella*, etc. Wear gloves if possible when handling raw meat.

Antagonistic muscles

Muscle is a good example of a tissue, not least because it is part of the structure of so many other organs such as the stomach and the eye. It is very important to help students grasp the idea that muscle tissue can only pull, as the protein fibres contract. Once this is clearly understood, the idea of antagonistic pairs of muscles moving parts of the skeleton makes sense. Students have a useful practical aid in their own muscles, particularly the biceps and triceps. This fits in well with standard textbook diagrams too.

Enhancement ideas

- Both smooth muscle found in the gut, and cardiac muscle in the heart can be introduced here. Details of structural differences are probably not needed but some groups may find it interesting to see the differences from slides projected on a screen, and to link differences in the structure to differences in the functioning of the muscles.

4.4 Exercise and fitness

Previous knowledge and experience

Students will have considered whether taking exercise helps to keep people healthy.

A teaching sequence

Plan your own desired learning outcomes for the students, to give you a tool for evaluating your teaching and your students' learning as you move through the topic. Giving students a broad overview of the work to come and asking questions to find out the level of prior knowledge of the group you are teaching – maybe in the form of a quiz, or a questionnaire, a traffic-lights exercise or simply questions as you talk – will set the scene for the topic and enable you to pitch it at the right level. This topic has lots of opportunity for individuals to use themselves as experimental organisms, and lots of intrinsic interest.

It is useful to begin by finding out what students understand by the term 'fitness'. They will have met the idea before in their primary schools and it is very interesting to find out what they have taken on board. A useful working idea – although you would not necessarily expect students to express it in quite these terms – is that a fit body is capable of carrying out all of the exercise demands of normal life without any excessive stress, and can cope with sudden extra demands when necessary.

Depending at what point you cover the work on fitness, you can use it as a good example of integrated body systems. Dealt with fully, it encompasses the cardiovascular system, the breathing system, the skeleton and muscles; even the senses and the gut can be mentioned.

■ The consequences of exercise

When you exercise, your muscles are working, and the more you use particular muscles the more your body will respond by building extra muscle in that area. This is why varied exercise results in a more toned and muscular body. Similarly, if you are ill and in bed for a week or two (e.g. as a result of flu) then your body breaks down muscles that are not being used and you get muscle wastage. This contributes to the weak and tired sensations you experience when you try to get up and about again – you have to take time to rebuild your muscles. A good example of this muscle wastage can be seen when people have broken a leg or an arm and had it encased in plaster for several weeks.

When you exercise and your muscles work, the demand for oxygen goes up. Practical work measuring the response of students' own bodies to exercise is a really useful teaching tool here, and there are a number of important ideas to get across. This work can be done in the laboratory, in a sports hall or outside, depending on the nature of the group, space available and your own choice.

It is very important before students undertake exercise to check if any of the class do not do PE for health reasons, or if individuals need to use asthma inhalers before exercising. Although they will not be doing anything too strenuous, these precautions should always be taken. Also, pre-warn students to bring in PE shoes to avoid any possible injuries. If there are several students who cannot exercise, use them to help – they can do the timing, record the breathing rates of small groups of students, etc., so that they are included in the practical.

1 It is necessary for each individual to have a fairly accurate picture of their own resting breathing rate. They must sit still and in silence for a few minutes, breathing normally, and then count the number of breaths they take over each of three 30-second periods. They must not move or talk during the measuring, just note down the number of breaths at each count. One breath counts as breathing in and then out again. Each of their results can then be doubled, and then the average of the three numbers found. This will give them their average breathing rate per minute. It is important to stress the need for them to be completely *at rest* when they are measuring their breathing, both at this stage and after exercise. This practical may give you one of the quietest lessons on record!

2 Students should then undertake a minute of gentle exercise and then, staying still and quiet, record their breathing rate at the end of that exercise and for each of 5 subsequent minutes, by when, for most students, it will have returned to the resting rate.

3 Students can then undertake a more vigorous minute's exercise, or a longer period of gentle exercise, and repeat the process of measuring their breathing rate for 5 minutes afterwards.

You may choose to let students develop their own hypotheses on the effects of exercise on breathing rate and then plan this experiment for themselves. Alternatively, you may ask them to follow instructions, or give instructions on the basic method but give them the freedom to decide how to vary the periods of exercise, by changing either the duration, the intensity or both.

A clear worksheet with a results table to fill in will help less able students to cope with this, and you may need to go through the explanation of finding the average breathing rate step-by-step on the board.

One of the nice things about this sort of experiment is that it provides students with real raw data which they can use in a number of ways. Students may simply work with their own data or the whole class might feed their results into a spreadsheet so that every student can have data from the whole class to work with (obviously the value of this will depend on whether everyone undertook similar exercise). This can provide some useful practical experience of How Science Works (HSW). When looking at the results and trying to draw overall conclusions, some of the problems of producing reliable data and then interpreting these data can be brought home to your students – as well the satisfaction of seeing patterns emerging.

■ Anaerobic respiration

This is a good place to introduce or revisit the idea of anaerobic respiration and oxygen debt. As exercise increases, more energy is needed and so respiration levels go up in the muscle cells. Often the need for extra oxygen is not met immediately, and in vigorous exercise it is sometimes impossible for the body to supply all the oxygen needed. In these cases anaerobic respiration takes place in the cells. Glucose is broken down without oxygen. This is inefficient – it doesn't produce very much energy and lactic acid is formed as a waste product. Lactic acid (strictly, lactate) builds up in the muscle, causing pain and eventually stopping them from working. Blood flowing through the muscle removes the lactic acid. When exercise stops, you will have built up an 'oxygen debt' during the time you have continued to work with insufficient oxygen. Although you are at rest, your body cells continue to need more oxygen than usual because oxygen is needed to get rid of the lactic acid. The oxygen debt is the amount of oxygen needed to break down this accumulated lactic acid. Only once the oxygen debt is paid off does your breathing (rate and depth) return to normal.

■ Fitness

When people exercise regularly they become fitter. They have bigger muscles with a good blood supply to provide oxygen and food to the respiring muscle cells and remove the waste products of respiration. Their hearts become more muscular and so stronger, capable of pumping more blood more rapidly than an unfit heart. Their lungs become bigger, with a better blood supply so that they are capable of breathing more air into their lungs with each breath and carrying more oxygen into the body from each breath. Increased fitness is linked with increased health – students can be given data both on the effects of exercise on health and on the numbers in the population who take part in regular exercise as a stimulus to discussion. This again contains good HSW material.

Another useful piece of research students can do is to use internet resources to help them identify why people who exercise more are healthier. There is clear evidence that people who exercise more have a lower body mass index (BMI) and are less likely to be overweight or obese than people who don't do much exercise. This means they are less likely to suffer the diseases associated with obesity – heart disease, high blood pressure and also type 2 diabetes. This offers a useful tie in with other areas of the specifications, particularly general material on keeping healthy.

Further activities

- Either in conjunction with the ideas below or as a free-standing exercise, students could design a poster/leaflet/advert for local radio aimed at informing young people of the benefits of exercise and encouraging them to get fitter.

Enhancement ideas

- Classroom Medics bring physiology into the classroom and give your students the chance to find out about changes in breathing and heart rates, both using a medical dummy and on themselves using real medical equipment. The team runs workshops that are both very informative and extremely motivating for students – they visit schools anywhere in the country. Find out more about what is offered at http://www.classroommedics.co.uk.

- Many leisure centres now run GP referral schemes where patients with cardiovascular and respiratory problems are prescribed courses of supervised exercise instead of drugs. Either you or your students may be able to talk to staff and even look at some of the results achieved. It is important to make sure that leisure centre staff are aware that you do not want to know who is involved in these schemes, simply the data on health benefits.
- Students will probably have seen adverts for 'fitness assessment' in their local leisure centre. Students can find out about the various measurements taken to assess fitness and think about the accuracy of these measures. You may find your local leisure centre willing to provide someone to come along and talk to students and demonstrate some of the tests. It is, after all, in their interest to convince young people of the need for exercise!

Other resources

ICT resources

There are official and unofficial ICT resources that can be used to enhance your teaching, support weaker students or stretch more able ones. The internet provides you with many data sources, e.g. the World Health Organization, which are changing and being updated all the time. You can find these at sites such as www.ase.org.uk, where the necessary shortlisting has been done.

Sites such as those listed below can be very useful for students to explore:

www.bbc.co.uk/science/humanbody/
www.medtropolis.com/VBody.asp

A number of resources relevant to curriculum work are recommended by David Sang and Roger Frost in *Teaching Secondary Science using ICT* (ASE John Murray Science Practice). These resources suggest encouraging students to carry out a survey of people's attitudes to health and fitness and then using a database program to analyse their data by sorting and graphing.

There are many different resources available online, which can be used to demonstrate fitness programmes, the effect of fitness on health, breathing rates, oxygen debt, etc. These resources tend to change frequently, so it is worth having a look each time before teaching a particular topic.

Students can learn a lot using sensors and data-logging technology. For example, the air in a container of worms or maggots can be analysed over time and changes in humidity and oxygen levels measured. An exciting variation on this practical is to allow students to re-breathe the air in a polythene bag while the changes in humidity and oxygen levels are measured. They should do this for 5–10 breaths ONLY. You can do a similar thing as a short demonstration if it does not seem appropriate to do it as a class experiment. Sensors can also be used to measure the effect of exercise on heart and breathing rates.

Be very careful to ensure that students only re-breathe the air in a bag for 5–10 breaths, with 10 as the maximum.

Science Learning Centres (www.sciencelearningcentres.org.uk) are full of ideas and resources to help you deliver this and other areas of biology as effectively as possible. The Society of Biology (www.societyofbiology.org) also provides many resources and pointers to help you find appropriate material for your students.

The Natural History Museum (www.nhm.ac.uk) often has excellent displays explaining how various systems of the human body work.

5 Communication and control

Mike Cassidy

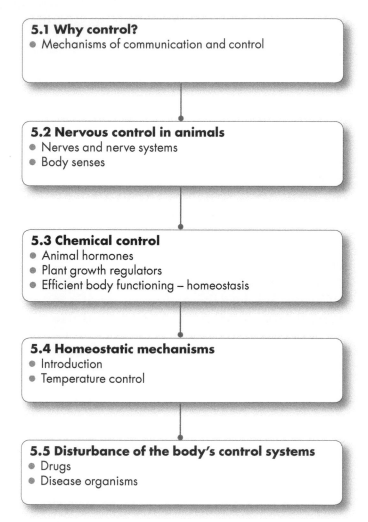

5.1 Why control?
- Mechanisms of communication and control

5.2 Nervous control in animals
- Nerves and nerve systems
- Body senses

5.3 Chemical control
- Animal hormones
- Plant growth regulators
- Efficient body functioning – homeostasis

5.4 Homeostatic mechanisms
- Introduction
- Temperature control

5.5 Disturbance of the body's control systems
- Drugs
- Disease organisms

Choosing a route

The topic of communication and control requires an overview of several functioning body systems and underlying principles. Teachers scaffold a variety of control mechanisms (nervous and chemical action) around a central core.

A teaching sequence might begin with the complexity of living organisms; either body complexity, for example the trillions of cells in the human body, or cell complexity.

Students can be introduced to complex machines (anything from washing machines to automobiles) and the need to integrate the component parts. A jumble of 'spare parts' cannot function; only when assembled correctly are they *potentially* capable of functioning. And there needs to be a stimulus (switching on, turning the ignition key) to start the process. All control systems, either animate or inanimate, operate along the same lines:

- Control mechanisms are centred about a stimulus–response model.
- A sensation or stimulus is detected and causes the organism to make a response.
- Nerves or hormones make up the means of communication between constituent parts.

Animals are complex, multicellular creatures with a requirement for relatively sophisticated control. Plants are also active organisms with comparable (though not identical) control requirements. This is less well understood by students. Plants are generally seen as inactive. This misapprehension can be corrected by fast-frame video images that show the life of plants (speeded up) as being equally vigorous and almost as complex.

Homeostasis is seen to be a particularly difficult topic by educators. It is introduced through specific and easily observable examples such as temperature control. Post-16 biology extends the range of examples (for example osmoregulation) and presents an underlying systems approach. The topics in this chapter are more likely to be encountered by older students.

In this section we will look at the *process* of co-ordination and the *mechanisms* of control and communication.

5.1 Why control?

Previous knowledge and experience

Students will know that the bodies of animals and plants are complex. They will almost certainly have encountered the analogy of comparing the human body with a machine. But the need for control (Section 5.1) and the nature of that control (Sections 5.2 and 5.3) need to be introduced through example and analogy.

Life processes will have been taught as having survival value.

A teaching sequence

■ Maintaining a steady state

The ability to control and co-ordinate the activities of the body (both on a cellular and macroscopic level) is a feature of all living things.

Characteristics of living things are commonly taught at the beginning of a biology programme of study. They are generally listed using the mnemonics Mrs Gren (movement, reproduction, sensitivity, growth, respiration, excretion and nutrition), or Mr Green (movement, reproduction, growth, respiration, excretion, excitability, nutrition) which again emphasises the central role of sensitivity or responsiveness in the biological sciences.

Living organisms exist in a 'steady state' with respect to their environment. They maintain their integrity by moving materials into their bodies to balance those leaving – we call this a dynamic equilibrium. A lump of rock, on the other hand, will generally remain inert. This is referred to as static equilibrium.

The question then arises, 'How do living things maintain a steady state?'.

■ Control systems

Mechanisms of regulation in living things show features in common with the regulation of machines. Both organisms and machines achieve stability by control. The science of control systems is called cybernetics. Communication is achieved either by chemical (hormones) or electrical (nerves) means.

The basic components of a control system are:

input → detector → regulator → effector → output

Stability is achieved by establishing a standard operating level (the norm) and thereafter correcting any deviation from this. The efficiency of the system is determined by the degree of deviation from the norm. For instance, we might set a room thermostat at 25 °C. The control system then attempts to maintain this temperature, turning radiators on when too cold and turning them off when too hot.

It can be useful to ask students to name the component parts of a room's heating system and to compare them with the body's heating and cooling system (see Table 5.1).

Table 5.1 Comparison of body temperature control with a room heating system

Component	Room heating	Body temperature control
Input	room temperature	body temperature
Detector	thermometer	specialised blood temperature receptors
Regulator	thermostat	brain (hypothalamus)
Effector	room heater	muscles (dilating or constricting superficial blood vessels and shivering response); sweat glands
Output	heat generated (or not)	heat generated, heat conserved or heat lost

Effective control systems rely on their components being linked together. Information flows from detector to regulator to effector. Communication between component parts is achieved by nerve impulses or hormones.

Feedback is used to inform the control system as to how effective these corrective mechanisms have been and whether further effort is needed to return to the norm. A feedback loop is built into most control systems, both living and non-living.

The basic components of an animal control system are analogous to those of a machine:

Machine: input → detector → regulator → effector → output
Animal: stimulus → receptor → central nervous system → effector → response

The components are linked in a specific way forming a stimulus–response (S–R) chain (Figure 5.1).

Figure 5.1 The stimulus–response chain

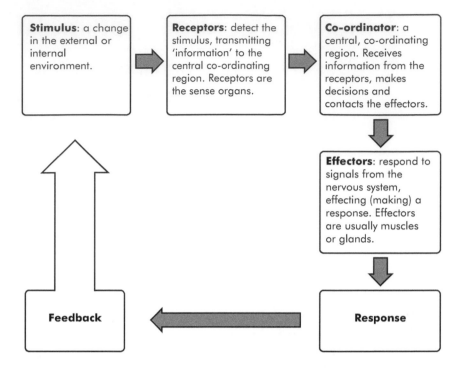

Further activities

- Reaction times can be measured by a student grasping a falling ruler (the distance the ruler falls is a measure of reaction time). Ask students to draw a flowchart representing the S–R chain.
- Several animal behaviours can be explored in the light of a stimulus–response chain. Freshwater flatworms, for example, are relatively easy to trap (a piece of liver in a jam jar) in shallow ponds and will show marked responses to food. Their movement in a Petri dish can be recorded before and after inserting a small piece of meat. The (chemical) stimulus causes a marked response as the animal quickly moves towards the food. Remember to return all animals, taken from outside the classroom, to their original habitat.

Enhancement ideas

- Video material (for example, *Stimulus Response*, Association for the Study of Animal Behaviour) of simple animal behaviours can be used to demonstrate S–R links.
- Students engage in role play, passing the 'message' (a parcel) between stimulus, control region and effector.

5.2 Nervous control in animals

Previous knowledge and experience

Students will have met a variety of body senses and carried out investigations into sight, hearing and taste. They may not be aware of internal sensations and internal responses.

An awareness of voluntary and involuntary actions is encouraged at an early age.

A teaching sequence

■ The nervous system

One of the characteristics of the animal kingdom is that all animals possess a nervous system. Even jellyfish have a simple nerve net. A nervous system is a collection of excitable cells arranged in a meaningful way.

If students are asked to name parts of the nervous system, responses such as 'brain' and 'nerves' are often obtained. Show students a diagram of the human nervous system. It is made up of two parts: (i) a central region containing brain and spinal cord together with (ii) associated peripheral nerves (emanating horizontally). The term 'central nervous system' is used to label this middle portion.

The nerve cell (or neurone) is specialised for conducting impulses. Like all body cells, neurones are microscopic but some (such as those coming out of the spinal column) can be very long, reaching over a metre in length. The change in electrical potential (due to movement of ions across the outer nerve cell membrane) is called an action potential.

Nerve cells:

- are often covered in a fatty sheath for (electrical) insulation
- are separated by gaps (or synapses) at their ends
- can make many connections
- can modify their connections (the basis of learning).

The speed of a nerve impulse can be estimated by having a group of students standing in a circle. One student 'passes on' a message by squeezing the hand of the individual to their left, who immediately does the same. Time taken to pass the message between students is recorded whilst distance taken (hand to brain to hand) is estimated by tracing the nerve pathway using string. Velocity can then be calculated using distance/time (in metres per second, ms^{-1}).

Body reflexes are another way of investigating the nervous system. A reflex is a rapid, involuntary response to a stimulus. The eye-blink reflex, kneejerk reflex and swallowing reflex are good examples (though the swallowing reflex is quite complex and the kneejerk reflex is unusual in not having a connector neurone).

Using an example, such as removing the hand from a hot object, the survival value of body reflexes can be seen.

Figure 5.2
Components of a reflex arc

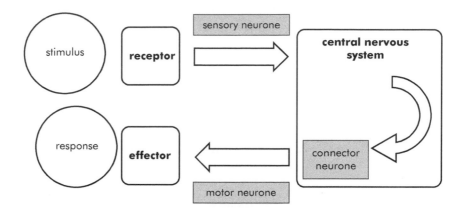

Figure 5.2 shows the three main types of nerve cell:

- sensory neurone – taking impulses from the receptor
- motor neurone – taking impulses to the effector
- connector neurone – joins the input component with the output.

NB: students are tempted to refer to neurones as 'nerves' (sensory nerve, etc.). Remind them that a neurone is a single cell, whereas a nerve is a collection of cells (or, more accurately, cell fibres).

■ Body senses

Discuss with students the responses made by animals to a variety of stimuli:

- bees visiting brightly coloured flowers (colour)
- sharks hunting by smell (molecules in water)
- houseflies landing on food and 'tasting' food with their feet (molecules on the surface).

A receptor is the part of the body that is adapted to receive stimuli. It detects stimuli both from the external and also the internal environments.

Essentially, a receptor works by the stimulus causing changes in the ionic balance of the nerve cell membrane. This then causes the production of a nerve impulse. Nerve impulses are:

- fast
- electrical
- generated with an 'all or nothing' response (no half measures; they fire or they don't)
- unidirectional (travel in one direction only).

We are generally thought to have five senses (sight, hearing, touch, taste, smell) but others have been identified, most notably proprioception (knowing where your body is in relation to itself) or sense of equilibrium.

There are many different kinds of receptors in the human body and they are classified in several ways (Table 5.2). Students usually do not need to know the names of specific receptors; simply that the body responds to many different kinds of stimulation, both internal and external.

Table 5.2 Classifying body receptors

Classified by:	Receptor name	Stimulus	Example
Their complexity	General senses	single cells or small groups of cells; respond to a variety of stimuli	pain sense endings, touch receptors, pressure receptors in blood vessels
	Special senses	complex sense organs; respond to specific stimuli	eye, ear
	Photoreceptor	light	eye
	Chemoreceptor	chemical	taste buds, receptors in the nose
Stimulus type	Thermoreceptor	heat	simple receptors in the skin
	Mechanoreceptor	mechanical (e.g. touch) stimuli	pressure receptors and touch receptors in the skin
	Baroreceptor	pressure	pressure receptors in blood vessels

■ The special senses: the eye

Eyes are complex structures adapted to receiving light and transmitting information to a central co-ordinating region (brain). Eyes range from simple eye cups in flatworms (that simply indicate light and shade and the direction light is coming from) to the complex vertebrate eye with structures for focusing light into sharp images and detecting different wavelengths or colours.

In vertical section the eye is seen to be made up of a spherical bag of jelly separated by a crystalline lens. Light is focused by the lens (it alters its shape due to the contraction of a muscle ring) on to the light-sensitive layer, called the retina, at the back of the eye. At the front of the eye (Figure 5.3) the muscular iris (the coloured part) contracts or relaxes, decreasing or increasing the size of the pupil (the black dot in the centre), thereby altering the amount of light entering the eye. Too much light and the retina can be irreparably damaged.

Figure 5.3 External (visible) features of the human eye

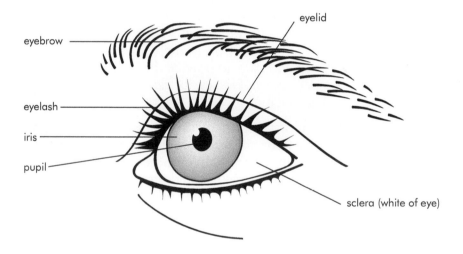

Students find difficulty in appreciating the eye's internal structure through diagrams alone. The suspension of the lens is a particularly troublesome area. Model eyes and eye dissection are particularly useful in this respect. Check your school policy on dissections.

Note that the optic nerve connects the eye to the visual centre of the brain (located at the back of the brain). The optic nerve is a large nerve. If the human skull is examined, a hole at the back of the eye socket indicates where the optic nerve leaves the eye on its way to the brain.

At a point just below the centre of the retina, the optic nerve collects together sensory fibres from retinal cells. There are no light-sensitive cells (rods and cones) here. So at this point light falling on the retina cannot be detected. We call this region the blind spot. If you draw a cross and a large dot 10 cm apart on a sheet of white paper, and hold it at arm's length, both symbols can be clearly seen. Focus on the cross and close your left eye. Bring the paper slowly towards the face and you will see the spot disappear. At this point the image of the cross is picked up by the rods and cones but the image of the spot is focused on the blind spot which has no light sensitive capability.

The physics of the eye may be studied with reference to work on lenses, refraction, reflection, wavelengths and energy transformation. An effective strategy is jointly to plan sessions with physics teaching staff in order to reinforce the basic principles of optics.

Vertebrate eyes (especially those of birds and mammals) are particularly good at 'seeing'. They:
- have a lens that can form an image
- can adjust the lens to focus on near and distant objects
- can generally control the amount of light falling on to the retina
- can often work in both low and high light intensities
- can generally distinguish different colours
- usually have sharp (acute) vision
- come in pairs to provide three-dimensional vision and depth perception. Horses, for example, have eyes on the side of their head which gives a different perception.

The retina is made up of light-sensitive cells. Cone cells provide us with colour vision and, because they are densely packed, give us especially acute (high resolution) vision. Rod cells work at lower light levels than cones and provide us with simple monochrome (black and white) vision. Ask students to consider what happens to the colour of trees, hedges and flowers as light falls – everything becomes more monochrome as only the rods work in low light intensities. Incidentally, rod cells are concentrated at the edges of the retina (that is at the periphery of our field of view). Hence sentries posted at night in army barracks, etc. learn to look at objects through the corners of their eyes. Students may also appreciate that we have different types of cone cell (responding to the three colour wavelengths: red, green and blue). Some students may not be able to discriminate between certain colours – we call this colour blindness – due to lack of a particular type of cone.

Further activities

- Ask students how they are able to see objects close to them as well as objects further away. If you were using a pair of binoculars or manual camera you would need to adjust the focus ring – that is, move the lens forwards or backwards. We cannot do this with the human eye. But we can alter the shape of each lens, and we do this to focus on objects at different distances. You can sometimes feel the 'pulling' of the radial fibres (rather like spokes of a wheel) altering the shape of the jelly-like lens. We call this ability accommodation. Find some 3D vision activities to do as well.

Enhancement ideas

- Photochemical reactions are commonplace, from the bleaching of coloured fabrics in sunlight to the action of light on silver nitrate (as used in old style, pre-digital, photographic films). So students should not be surprised to learn that light falling on pigments inside receptor cells (rods and cones) causes these pigments to break down (a reversible reaction, of course) thereby causing an action potential and releasing a nerve impulse.
- The eye responds to light stimuli, but the eye does not 'see'. It is the brain that makes sense of the information from the eye and it is therefore the brain that sees. Students are often excited to view visual illusions. There are many of these to choose from and all illustrate the principle that our perception (that is the brain) can be tricked into seeing things that are not there.
- What is the function of (i) eyebrows, (ii) eyelids, (iii) tears?
- Find out about health issues that affect eyes such as cataracts, glaucoma and diabetes.

■ The special senses: the ear

The human ear has a dual function as both an organ of hearing and an organ of balance. It has a three-part structure:

1 outer ear – comprising a twisted ear canal and ending in the tympanum or eardrum
2 middle ear – comprising three small bones (ossicles) that transmit vibrations across this air-filled space
3 inner ear – a complex, fluid-filled region made up of an upper 'looped' region controlling our sense of balance and a lower coiled region where vibrations in the fluid are detected by sensory cells.

These structures are very small and they are best illustrated by use of models, posters or diagrams in textbooks. Most of the structures have both common and scientific names (e.g. eardrum/tympanum, ear lobe/pinna). Younger students use the common names; older (examination) groups use the scientific terms.

Hearing may be best approached through revision of the physics of sound. Noise comprises waves of compressed air (sound waves) created by a vibrating object. Reference to guitar strings, drums and 'twanging' rulers can be introduced here. Properties of sound include amplitude of the sound wave (loudness) and frequency of the sound wave (pitch, high notes/low notes). Sound waves can be illustrated visually by coloured images found on most amplifiers, MP3 players or music computer programs (and also the oscilloscope found in the physics department).

How we hear is complex, but in essence relates to detection of sound waves. These:

1 enter the ear, collected by the ear lobe (reference to ear trumpet?)
2 are channelled down the ear canal
3 cause sympathetic vibration of the eardrum
4 ensure vibrations are transmitted through the middle ear via three small bones (ear ossicles) touching both the eardrum and the inner oval window.

Then:

5 vibrations are transmitted through the fluid of the coiled cochlea from the oval window
6 sensory cells lining the cochlea respond to loudness (how much these small cells are displaced) and pitch (different patches of cells responding to different note frequencies)
7 the sound wave is dampened by hitting the round window (a membrane at the far end of the cochlea).

How we sense position and balance is achieved by:

- movement of fluid in the 'looped' region, the semicircular canals (together with the sac-like region below) affecting patches of sensory cells
- the direction in which our head is tilted.

Students can plot the routes of sound waves on to pre-prepared diagrams.

Further activities

- Hearing can be studied indirectly by asking students to respond to various sound stimuli, for example:
  - A signal frequency generator (borrowed from the physics department) connected to a loudspeaker can generate sounds of varying frequency. Young people can generally hear notes of between 20 and 20 000 Hz (cycles per second). Older teachers have reduced frequency discrimination.
  - Sensitivity to sound can be determined by holding a ticking watch at varying distances from a blindfolded subject.

- Make sure that students know about the health issues of very loud noise.
- Ability to sense the direction of sound can also be determined using a blindfold, a subject and a ticking watch.
- Use of an ear trumpet (made from thin card) is seen to increase the sensitivity to both sound and directional ability. Why is this?
- Use of a tuning fork can demonstrate how vibrations can travel through both air and water. Talking into an inflated balloon (feel the vibrations) is used to demonstrate this feature to children with hearing difficulties.

Balance can be explored as follows:

- Have students stand still for 3 minutes (they will sway slightly). Notice how they correct their balance. With eyes closed, students find this task much more difficult (no reference point to focus on).
- A model of the ear's three semicircular canals can be made from stoppered, clear plastic tubing. The three semicircular canals can be constructed separately but attached to each other at right angles – as would be the case in real life. Trapped air bubbles will demonstrate displacement of the fluid as the head is tilted from side to side.

Enhancement ideas
- Ask students to consider the consequences of deafness in an individual. The RNID (Royal National Institute for the Deaf) can help with case histories.

■ The general senses (touch, taste and smell)

The skin is the largest organ in the body and home to several senses – pain, touch, pressure and heat. Taste and smell are chemical senses located in the human tongue and nose respectively. All are known as general senses and detection is brought about by a relatively small cluster of cells.

Little detail is required in current frameworks and specifications but some misconceptions may be addressed here:

- What we think we are 'tasting' (e.g. the taste of an onion) is actually a combination of taste and smell. When wearing a nose clip an onion tastes sweet! Students can rarely tell the difference between a slice of apple and a slice of onion gently placed on the tongue (providing a nose clip is in place, and there is no looking and no chewing).
- The little 'bumps' on the tongue are not taste buds. Rather, they are papillae, designed to roughen the tongue's surface. (Just

think how difficult it would be to chew and swallow with a smooth, shiny tongue!) Ask students if they have been licked by a cat or a large herbivore such as a cow.

- Our skin is not equally sensitive over the body. Using a small piece of card with two pins (placed 1 cm apart) ask students to test different skin areas for sensitivity. This is achieved by students working in pairs, with the experimenter applying either one or two pins to the skin surface and the subject responding (can they recognise if they are being touched by one pin or two?). After 20 presentations the number of correct responses is recorded. Which areas do they think are more sensitive – fingertip, back of hand, back of neck, lips?
- There are now thought to be five taste sensations: the original sweet, salt, sour and bitter along with umami (a sort of savoury/meaty flavour).
- The classic taste map of the tongue (where different tastes are located in different regions of the tongue surface) is now known to be invalid. Recent research has shown that there is sensitivity to taste across all regions of the tongue.

PROCEDURE

The classic skin temperature investigation is to use three beakers of water: (i) cold water at 2 °C, (ii) lukewarm water at room temperature, 20 °C and (iii) hot water at 40 °C. The index finger of the left hand is placed in the cold water for 2 minutes whilst that of the right hand is placed in the hot water. After this 2-minute acclimatisation, both fingers are placed into the central beaker of lukewarm water. Ask the students what they experience. The finger that was in cold water will feel warm whilst that in the hot water will feel cold. This is because skin temperature receptors are not tiny thermometers registering absolute temperature. Rather they record a change in temperature above or below a 'thermoneutral' temperature of around 20 °C.

Further activities

- There is still debate about the identity of temperature receptors in human skin. But it is thought that specialised free nerve endings register any change in temperature. A temperature above 45 °C or below 10 °C is also experienced as pain.
- The classic skin temperature investigation described above may be extended by placing the bulb of a thermometer in each of the three beakers, wiping them dry and touching the skin at various points. Do you sense only touch or is temperature sensed?

Enhancement ideas

- Discuss why a heavy cold can cause difficulty in detecting odours (excessive mucous production in the nasal cavity interferes with the functioning of the olfactory (smell) receptors).

- Which animals rely primarily on their sense of smell? And why should this be?
- A recent study (*Scientific American*, April 2009) found that human females generally have a better developed sense of smell than males. How might you investigate this in the classroom?

5.3 Chemical control

Chemical communication is universal in animals, plants and microbes. Chemical signals, though, are usually relatively slow and diffuse. They work well over small distances (as in cell-to-cell signalling) or for slow, measured responses such as growth and reproductive cycles. Chemicals, though, cannot mediate fast responses such as reflexes; they are also persistent and therefore need to be broken down.

Previous knowledge and experience

Students may be familiar with specific examples of hormone action (insulin to lower blood sugar, sex hormones to control puberty and reproduction) but are less likely to have an integrated picture of the role chemical signals play in control and co-ordination.

At an earlier stage in their development, students will have studied plant germination and conditions needed for plant growth. They will have come across animal hormones and may be aware of rooting powders, which are plant growth regulators.

Students can review the names of plant parts together with the process of germination.

A teaching sequence

■ Animal hormones

Animals possess two systems of internal communication and control:

1 (electrical) nervous system
2 (chemical) endocrine (hormone) system.

A hormone is a secretion, released into the bloodstream, which has an effect on a distant structure (target organ).

A gland is any body structure that produces a secretion. An endocrine gland does not possess a tube or duct to release its secretion (unlike, say, a sweat gland). The secretion, a hormone, is released directly into the bloodstream.

A hormone works by specifically affecting the outer membranes of its target cells. This generally sets up a chain of biochemical reactions within the cell, causing the alteration of its chemistry together with the production, and maybe release, of a cell-produced compound.

Hormones are involved in three main areas of physiological function:

1 growth and development
2 reproduction
3 maintenance of the internal environment (homeostasis).

Examples of hormone action for several of these functions are provided in Table 5.3.

Table 5.3 A summary of major hormones and their effects

Endocrine gland	Hormone produced	Effects
pancreas	insulin	lowers blood sugar
	glucagon	raises blood sugar
adrenal gland	adrenaline	gets the body 'ready for action' by raising blood sugar and increasing chemical activities and general awareness
	cortisol	helps the body resist stress by raising blood sugar
thyroid gland	thyroxine	increases the body's general metabolic rate (causes metamorphosis in frogs)
sex organs	testosterone (produced by the male testes)	promotes formation of male secondary characteristics including sperm formation; involved in male courtship behaviour
	oestrogen (produced by the female ovary)	promotes development of female secondary sexual characteristics; stimulates growth of the uterine lining during early parts of the menstrual cycle
	progesterone (produced by the female ovary during the menstrual cycle)	completes the development of the uterine lining and maintains this lining if fertilisation takes place with an embryo implanted there
pituitary gland	growth hormone	increases growth rate of young animal
	thyroid stimulating hormone (TSH)	stimulates hormone production in the thyroid gland
	follicle stimulating hormone (FSH)	stimulates production of follicles within the ovary (resulting in the shedding of eggs)
	anti-diuretic hormone (ADH)	stimulates water reabsorption in the kidneys in times of water deficit
	oxytocin	stimulates uterus muscles during childbirth and milk release during suckling

Specific information can be introduced at the relevant syllabus point:

- Reproductive hormones can be addressed when discussing growth and development, the menstrual cycle and birth.
- Control of blood sugar is explored when discussing transport of the products of digestion (although this is a popular example when describing principles of homeostasis).
- The action of the pituitary growth hormone is included when describing patterns of vertebrate growth and development.

However, the pituitary gland has such an important co-ordinating role (it affects other endocrine glands) that it should be treated separately.

Students should be encouraged to locate and identify endocrine glands on an outline diagram of the human body.

Further activities

- You might wish to explore a case study of a patient suffering from diabetes, describing both the symptoms and the mechanism of insulin treatment.

- Ensure that you do not embarrass students in your class that may suffer from this condition or have relatives who do.

A comparison of endocrine and nervous function can be carried out (see Table 5.4).

Table 5.4 A comparison of nervous and hormonal function

Factor	Nervous response	Endocrine response
speed of effect	generally very fast (fraction of a second)	relatively slow (hours, days, weeks)
area of effect	localised effect	general effect around an organ or around the body
timing	short-lived effects	long-term effects
blood supply	limited	very good capillary supply

Several questions may accompany the information in Table 5.4:

- Why are nerves much faster than hormones in achieving their effects? (Hormones have to travel in the bloodstream, whereas a nerve action potential moves rapidly along a nerve fibre.)
- Are there any fast acting hormones? (Yes: adrenaline, which acts in seconds.)
- If hormones have long-term effects, what happens to excess hormones in the bloodstream? (They are broken down after a period of time.)

- What types of body response are controlled by hormones rather than nerves? (Slow, moderated and controlled responses such as growth, puberty and the menstrual cycle. These often require a gradual build up of hormone.)

Enhancement ideas

- Pheromones are sometimes called 'external hormones' and are important in several areas of animal behaviour; for example receptive female mammals in some species release a scent to attract males and the queen bee releases a chemical to control the activities of the hive. In humans it is said that mothers can recognise their own offspring by scent alone. What might be the benefits in other mammals of recognising family members?

■ Plant growth regulators (plant hormones)

The term 'hormone' (from the Greek meaning to stir up or excite) is normally reserved for those secretions produced by *animal* endocrine glands. Substances that influence growth and development in plants are: (i) not produced by specific glandular structures; (ii) do not travel via a fluid flow. Plant biologists generally use the term 'plant growth regulators' or 'plant growth substances' (*Biological Nomenclature*, Society of Biology, 2010).

Plant responses to their environment can be observed by growing plants on window ledges (they bend towards the light), by growing plants from seeds (the shoot always grows upwards, the roots down, irrespective of how the seed is planted) and by observation (ivy clinging to walls, pea plants growing along supporting wires).

It is better that students observe these responses themselves. The plant 'behaviour' can be named:

- phototropism – a response to light (plants on the window ledge)
- geotropism – a response to gravity (growth of seedlings)
- thigmotropism – a response to touch (ivy growing close to walls).

Note that a 'tropism' is a growth response. Plants do not have muscles; the only way they can respond is by *growing* towards (positive tropism) or away from (negative tropism) a stimulus. If you examine a plant growing towards the light you can see that greatest curvature (that is, greatest growth) occurs on the shaded side. Does light therefore inhibit cell growth and cell division? What happens when plants are grown in the dark?

In the classroom, plant phototropisms can be demonstrated by growing mustard cress seeds on damp cotton wool in a small dish and exposing the seedlings (they germinate in 2 or 3 days) to:

(i) unidirectional light; (ii) light all around; (iii) no light. Cardboard boxes can be used to exclude all light or (if openings are made at one end) to provide unidirectional light. Seedlings outside the boxes are bathed in light all around. Small Petri dishes are useful for growing cress seeds. Some good video sequences on plant growth are available to show the class.

The question arises, 'What are the plant growth substances?' The answer is that they are often rather simple compounds produced by cells in particular regions (often the growing points). Some of the major plant growth substances and their effects are given in Table 5.5.

Table 5.5 Roles of main plant growth substances

Hormone	Effects	Examples
auxins	cause cell elongation and growth of stem and root	phototropism, cress bends towards light; used in hormone rooting powders
	high concentrations disrupt plant growth	synthetic auxins are used as weed killers
ethene	causes leaf ageing	important in leaf fall from certain trees
	ripens fruit	used commercially to ripen fruits such as lemons
abscisic acid	controls bud dormancy and is involved in leaf fall (abscission)	can transform the growing tip of birch trees into dormant buds; used to speed up leaf fall

Further activities

- Rapid cycling Brassicas (or Fast Plants) can be used to demonstrate plant responses to hormones. These plants may be purchased (along with planting instructions and worksheets) from SAPS (Science and Plants for Schools), Cambridge.
- Comparing plant and animal responses is an especially useful exercise (Table 5.6).

Table 5.6 A comparison of animal and plant responses

Animal responses	Plant responses
usually rapid	usually slow
short stimulus needed	prolonged stimulus needed
effect normally temporary	effect usually permanent
behavioural responses involve movement	growth responses are produced

Enhancement ideas

- Garden centres are useful suppliers of selective weed killers, hormone rooting powders and the like. Commercial uses of plant growth substances may be investigated as a project, using product labelling and classroom investigations (e.g. use of rooting powders on plant cuttings).

5.4 Homeostatic mechanisms

Previous knowledge and experience

Students may have encountered examples of control systems (such as temperature control) whilst studying 'Life processes and living things'.

There may be rudimentary knowledge of nerve action.

A teaching sequence

■ Introduction to homeostasis

When observing blood pressure or body temperature it is significant that humans (and other animals) maintain a constancy of internal factors such as temperature, electrolyte balance, etc. Irrespective of external conditions, animals strive to maintain constant conditions internally. Claude Bernard (1813–1878), a French physiology professor, was one of the first individuals to recognise and state this principle of internal constancy.

The more we understand cell biology, and cell chemistry in particular, the more we realise that individual body cells are vulnerable to even slight changes in conditions. Protein molecules, at the heart of much of our biochemistry, are readily altered by temperature, whilst the outer cell membrane can be easily damaged by osmotic changes. Thus, a constant internal environment is a prerequisite of continued cell existence.

Homeostasis is an ability of organisms to maintain chemical equilibrium (Figure 5.4). It involves self-adjusting mechanisms that return bodily function to a norm or set point. Therefore, in order to maintain water balance, receptors measure the viscosity ('stickiness') of the blood. Dilute body fluids (too much water) triggers a response leading to loss of water whilst concentrated body fluids cause water retention and thirst.

Figure 5.4 Use of correction and feedback in maintaining a constant internal environment

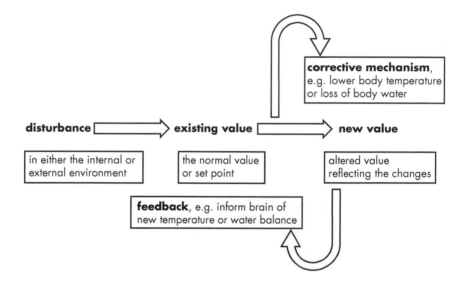

■ Temperature control in animals

Temperature control (thermoregulation) is commonly used to illustrate the principles of homeostasis to students.

Students will be familiar with the terms 'warm-blooded' and 'cold-blooded' even though these phrases are incorrect (the blood of a tropical fish is quite warm). The intention is to distinguish different types of *mechanism* used in regulating body temperature.

In humans, several methods are used to raise body temperature:

- increase in heat production – through raising the metabolic rate of the liver and other organs, active muscle contraction – shivering
- thermal insulation – to maintain existing body heat, body fat and addition of extra layers of clothing
- reduction of active cooling – reduction in sweat production, constriction of blood vessels near the skin surface.

Inverse mechanisms are used to lower body temperature:

- decreased heat production – lowering the metabolic rate of organs
- decreased thermal insulation – removal of clothing
- increase in active cooling mechanisms – increased sweat production, dilation of blood vessels near the skin surface; active 'flapping' to cool the body down.

Mammals (including humans) and birds that are able to use bodily or physiological methods to control their temperature are referred to as homeotherms (see Table 5.7). Animals ('lower' vertebrates and invertebrates) that lack active physiological mechanisms such as sweating and shivering are known as poikilotherms.

Poikilotherms (also called ectotherms because they generate heat from outside not from within) still manage to regulate their body temperature but they do this by behavioural methods rather than physiological ones.

Figure 5.5

Table 5.7 Temperature control in a small mammal (homeothermic)

External temperature low	External temperature high
narrowing of blood vessels near skin surface to conserve heat	widening of blood vessels near skin surface to lose excess heat
small (erector) muscles in skin cause hairs to stand on end, trapping a layer of air	hairs flatten against the skin
sweat glands produce very little sweat	sweat glands produce excess sweat, which evaporates from the skin surface
general increase in metabolic rate (non-shivering heat response)	
shivering of voluntary muscle (high energy cost, not efficient for long periods)	
behavioural responses such as moving to a warmer spot	behavioural responses such as moving to a cooler position
result = warming of the body	result = cooling of the body

Figure 5.6

Table 5.8 Temperature control in a lizard (poikilothermic)

External temperature low	External temperature high
aligns its body at right angles to the Sun to catch maximum rays from the Sun	aligns its body parallel to the Sun's rays to reduce the amount of sunlight
change to a darker skin colour to absorb more heat	change to a lighter skin colour to reflect more heat
	opening of mouth (thermal gaping) to lose heat by evaporation
	burrowing behaviour to avoid high surface temperatures
result = warming of the body	result = cooling of the body

Further activities

- Different sized beakers can be used to compare drop in temperature (draw cooling curves) with heat loss. A thermometer records drop in temperature whilst heat loss can be calculated knowing both the volume of water and the temperature drop.

  heat loss (measured in joules) =
  drop in temperature (°C) × volume of water (cm³) × 4.2

- Note that the smallest beaker has the greatest drop in temperature but the largest beaker has the greatest heat loss. Why is this? What significance does this have for large and small mammals?
- Students should be able to describe changes during exercise and their ability to lose heat generated by the contracting muscles.

Enhancement ideas

- Link the topic of temperature control to the physics of heat loss:
  - conduction
  - convection
  - radiation
  - evaporation.

What are the symptoms of hypothermia (low body temperature) and how might this condition be avoided in young people hiking in mountainous areas in cold, wet weather?

5.5 Disturbance of the body's control systems

The body is a sophisticated assembly of interacting parts. Homeostatic mechanisms regulate its working between fine parameters of temperature, water balance, pH, etc. However, modern human life exposes individuals to a wide range of substances they might not normally encounter. How the body reacts to these outside agents is the focus of this section. Of course, disturbances can also result from disease (e.g. diabetes).

Previous knowledge and experience

Students will have some knowledge of the effects of smoking tobacco, together with an awareness of alcohol and drug abuse from PSHE studies in school. Students gain as much knowledge on these products from friends, family and the wider media as they do from school. This is a fact we must not overlook. It is probable that their knowledge is detailed in parts but sketchy overall.

Although students will have information on infectious diseases it will be necessary to place the topic into a biological context.

A teaching sequence

This is a topic of great interest to young people and can be taught both in a social (science in society) and scientific context. Group activities and discussion work well. This topic also provides scope for argument and debate, perhaps exploring scientific models of drug action.

■ Use of drugs

A drug is a compound that alters the state of the body. Drugs provide little or no nutritional value and are taken either to benefit health (medicinal drugs) or to stimulate the body artificially (both legal and illegal drugs).

The former (beneficial) category, sometimes referred to as medicines, include:

- quinine – obtained from the quinine tree; prevents malaria
- morphine – obtained from the opium poppy; pain relief
- digitalin – obtained from the foxglove plant; used in heart medication.

The latter category includes:

- nicotine – obtained from tobacco leaves; stimulates heart rate, contributes to the build-up of fatty acids, causes physical and psychological dependence
- cannabis – obtained from the plant *Cannabis sativa*; a central nervous system depressant, hallucinogenic, can result in mental problems such as paranoia
- cocaine – made from the leaves of the South American coca plant; stimulates the nervous system, rapidly raises blood pressure, high physical and psychological dependence in a matter of days.

The word 'drugs' is often taken to mean 'illegal substances' but a technical vocabulary is necessary:

- A drug alters the chemistry, physiology or behaviour of a person.
- Drugs can be legal (prescription only or over-the-counter).
- Drugs may be illegal (class A, B or C) as defined by the Misuse of Drugs Act 1971.
- Drug misuse is taking the drug for the purpose for which it was made, but taking it improperly with a high dose, etc.
- Drug abuse is the deliberate use of a drug other than for its intended purpose.

When drugs are misused (e.g. excessive alcohol consumption) or abused (e.g. use of illegal drugs) by someone then that person's health is compromised. An individual can develop chemical dependence:

- Physical dependence – drugs such as barbiturates and heroin become part of the body chemistry; physical withdrawal symptoms are experienced when they are removed.
- Psychological dependence – is a craving (sometimes overwhelming) for the substance.

Further activities

- Students can interpret drug labels, for example those of soluble aspirin and vitamin tablets. Labels can be copied and students instructed to make data tables providing information such as:
  - active ingredients
  - % RDA (for vitamins)
  - purpose of the medication
  - dosage (age dependent?)
  - how to take the medication
  - possible side effects

- contraindications (when not to take the medication)
- expiry date
- storage conditions.
- Several commonly encountered compounds, such as caffeine, tobacco and alcohol, are available for study. Consider the following research topic: 'Alcohol is often viewed as a 'pick-me-up' or stimulant, whereas in fact it is a depressant. Examine the chemical effects of alcohol on the human body and consider the social effects of alcoholism.'

Enhancement ideas

- Students could consider the effects of drugs, for example, the social cost of illegal drugs or the medical effects on an individual.
- Ask students to identify decaffeinated food and drink items in their local food store. Students can then research the process of decaffeination. They can try to discover why some consumers prefer a decaffeinated product.

■ The introduction of disease organisms

An infectious disease promotes a reaction because the organism disrupts normal body functioning. The body reacts by attempting to neutralise the effects of the pathogen (harmful microbe). Common symptoms, such as fever, lethargy and loss of appetite, result; these can also have useful, adaptive functions and so aid recovery.

Agents of disease include bacteria, viruses and fungi. Most bacteria cause disease by releasing chemicals called toxins. These chemicals upset the chemical equilibrium of the infected host, causing familiar symptoms.

Further activities

- Students can explore body reactions to disease or body reactions to drugs and describe these in terms of disturbing the homeostatic balance of the body. How are these conditions managed?

Enhancement ideas

- Some researchers consider the entire planet as a self-regulating system. A theory of global homeostasis has been proposed with any disturbance being regulated by physical means (for example an increase in the Earth's temperature causes more evaporation therefore greater cloud cover therefore less heat energy reaching the surface and hence a cooling of the planet). Using their knowledge of ecology and bodily homeostasis, students could discuss the concept of a global homeostasis.

Other resources

Books

A comprehensive summary of 'Integration and Control' in the Animal Kingdom is found in: Jurd, R.D. (2004). *Instant Notes, Animal Biology, 2nd edition*. Oxford, Garland Science.

A clear review of human physiology, suitable for 16 year olds, is provided by: Wright, D. (2000). *Human Physiology and Health*. Oxford, Heinemann Education.

Teaching aids

Seeds, growing kits and practical activities can be found on the Science and Plants for Schools website www.saps.plantsci.cam.ac.uk.

Data-logging companies, such as Data Harvest, www.dataharvest. co.uk, provide specialist sensors and control systems for education (often with a full range of worksheets).

Video

Stimulus Response, a DVD produced for the Association for the Study of Animal Behaviour, http://asab.nottingham.ac.uk, considers the S–R chain in both humans and farm animals in sufficient detail for most secondary school work. Animal welfare issues are also addressed.

Websites

Teachers TV, www.teachers.tv, has a series of five videos on 'The Virtual Body' including teaching materials on 'Homeostasis' and 'Sensory systems'.

The American Physiology Society, www.theaps.org/education/ k12curric/index.asp, has internet teaching resources including 'The Physiology of Taste', 'It Takes a Lot of Nerve' and 'Brain Comparisons of Animals'.

The Association for the Study of Animal Behaviour, www.asab. nottingham.ac.uk, has a number of resources for teachers, including exercises, practical work and research information.

Background reading

The Association for Science Education's publication, *School Science Review*, (www.ase.org.uk) contains short 'Science Notes' on a variety of topics. Those of relevance here include:

- Thomason, B. (1992). Plant sensitivity, a historical source for teaching. *School Science Review*, **73** (264), pp. 97–101.
- Mackean, D. (1996). A cheap and cheerful choice chamber. *School Science Review*, **77** (281), pp. 70–71.
- Grant, P. (2006). A model of the ear's central canal. *School Science Review*, **88** (322), p. 11.
- Butler, K.G. (2000). Demonstrating hydrotropism in the roots of mustard cress or cress seedlings. *School Science Review*, **82** (299), pp. 95–96.
- Savoy, L.G. (1991). Valine inhibition of *E. coli* K-12: A simple demonstration of homeostasis. *School Science Review*, **72** (261), pp. 81–84.

Reproduction and sex education

Jennifer Harrison

6.1 Asexual reproduction in plants and animals
- Examples of asexual reproduction
- Mitosis
- Cloning and tissue culture
- Cancer

6.2 Characteristics of sexual reproduction in plants and animals
- Examples of plant sexual reproduction
- Examples of animal sexual reproduction
- Specialised cells; external or internal fertilisation; courtship
- External versus internal development; aftercare

6.3 Human reproduction and sex education
- Legal framework for Sex and Relationship Education (SRE)
- Physical and emotional changes at puberty and during adolescence
- Male and female reproductive systems: structures and functions
- Gamete production; sexual intercourse and fertilisation
- Menstrual cycle and control of fertility; contraception
- Sexually transmitted infections

6.4 Human pregnancy
- Fetal development
- Placenta
- Birth

Choosing a route

Reproduction is a crucial phase in the life cycle of any organism. It is the way in which a species is perpetuated, given that individuals must eventually die. Most students know that all living organisms grow and reproduce. However, they may have had little observational experience of reproduction in simpler organisms, and may have only limited understanding of reproductive processes in more complex ones. They may have limited understanding of asexual reproduction – the simplest form of reproduction – which does not involve sex, or sex cells. There may be confusion between reproduction (in terms of production of new cells for a subsequent generation) and growth (in terms of increase in number of cells, and therefore of mass and possibly of size of organism, for the same generation). A key difference between asexual and sexual reproduction is that, crucially, the latter results in new variations amongst the offspring. The connection between sexual reproduction and the process of meiosis (see Chapter 1) is frequently missing in students' understanding of some advantages of sexual reproduction.

The current science National Curriculum for primary school students covers key ideas in sexual reproduction and life cycle of flowering plants: fruits and seeds, seed dispersal, seed germination, insect pollination and flower parts for reproduction. It also includes human growth and development, as well as asking the important question, 'What is reproduction for?'

Students' earlier experiences of practical work with plant materials, such as flower structures, can be developed further. Extending students' knowledge of different types of seed dispersal (to include wind, water and animal dispersal) could also be explored. This would allow their early understandings of pollination, germination of seeds and seed dispersal to be checked. Students frequently confuse seed dispersal with types of pollination. This clarification can be followed up along with more detailed work on the reproduction of a specific animal, such as a frog or fish.

Care will be needed in planning for practical work with respect to the seasons and availability of specimens. Observation of the cycles of reproduction in plants generally requires long-term planning. Find out how many of your students have had the opportunity of growing and propagating plants or, indeed, of examining in detail a variety of living organisms. The topic as a whole provides abundant opportunities for bringing living things into your laboratory or classroom and providing students with hands-on experience. A well-stocked school greenhouse or pond can provide access to a variety of seasonal plant species.

Animal sexual reproduction presents good opportunities for maintaining and observing a variety of organisms throughout their life cycle; some organisms need little laboratory space and may have a life cycle that can be watched over just two weeks and independently of the seasons (e.g. fruit fly). Other organisms may demand more extensive maintenance, space and equipment, requiring observation over many weeks or months, and within specific breeding seasons.

The whole topic links directly with 'growth and development' in plants and animals. Basic microscopy skills are assumed, and students' ideas to do with microscopic size and the concept of magnification can be reinforced and extended. It is useful for students to know how to estimate the size of the cells they are examining (see Chapter 1).

Finally, your students need a firm understanding of the basics of nuclear and cell divisions, which lead to the formation of *either* genetically identical cells (as a result of mitosis) *or* gametes (as a result of meiosis) (see Chapter 1). They also need a grasp of the role of the genetic material DNA in the control of the cell's activities by the nucleus (see Chapters 1 and 7).

Teaching about human reproduction presents particular challenges to all teachers. Sex education in schools includes the study of human reproduction together with sexuality. Schools vary in their delivery of this area, in some cases deploying a well-integrated whole-school approach; in other cases devolving much of the area of study to the science department. It is crucial, as a teacher of biology, that you are well informed by your school and its policy for sex education about the precise part that you, as a science teacher, are to play in this sensitive area of education. You need to be properly briefed on any legal implications of this policy. Science teachers not only need to provide accurate information about reproduction and sexual health, but also to have strategies ready to deal with any awkward questions that might arise.

A particular responsibility of the science teacher is for aspects of health education. The current science National Curriculum is heavily focused on human biology and recognises the potential interest that students have in their health, as well as the importance for the whole school community of health-related issues. Bear in mind that science teachers are usually expected to have a more thorough knowledge of health-related issues than other teachers and are more likely to play a major role in the school's overall health education policy.

6.1 Asexual reproduction in plants and animals

Previous knowledge and experience

Students are likely to have met aspects of growth and reproduction in plants and animals that will have been mostly concerned with sexual rather than asexual reproduction.

A teaching sequence

Using a variety of examples, teachers can show students that asexual reproduction has several advantages over sexual reproduction:

- no energy is wasted in finding a mate
- many offspring are produced very rapidly
- favourable circumstances can be exploited very efficiently.

■ Examples of asexual reproduction

A variety of organisms can be used to show that there is only one parent in asexual reproduction. In each case some part of this parent divides to produce an identical individual, which then separates from the parent. A circus of activities (suggestions are largely seasonal) can provide an overview of the range of mechanisms for asexual reproduction in plants and animals:

- Estimation of the number of plantlets associated with a spider plant or *Bryophyllum*: ask students to suggest reasons for the production of so many.
- Illustration of reproduction (fission) in a bacterium or a unicellular organism, such as *Amoeba proteus*: use a sequence from a video or CD-ROM.
- Bread-mould cultures: observe the development of colonies of mould (a fungus) and the tiny black dots (sporangia) containing the spores. Ask students to suggest how these may be dispersed. Some fungal spores can trigger asthmatic and other allergic responses, so do keep cultures in closed plastic bags.

■ Mitosis

Build on your students' basic knowledge of this type of cell division. Explore with them the role of this division in that it allows cells (and therefore whole organisms) to reproduce themselves to form genetically identical offspring ('daughter cells'). You can extend this to make links with the role of mitosis in allowing multicellular organisms to grow and to repair themselves. Mitosis should eventually be understood as the basis of asexual reproduction (see Chapter 1).

■ Cloning and tissue culture

Asexual reproduction provides a genetically identical and stable gene pool in which a clone (e.g. a clump of daffodils) results from a single cell. You can consider examples from horticulture and agriculture (e.g. fruit trees and pineapples) where the advantage is that particular characteristics of the crop plant can be maintained from one generation to the next. Consider also how a change in an aspect of the environment, e.g. prolonged drought, can be a disadvantage to a species that is reliant on asexual reproduction. Notice also how a range of wild plants (e.g. grasses) use both methods of reproduction, so that the disadvantages of cloning become less significant (see Chapter 8).

The term 'clone' is also used in connection with genetic engineering (see Chapter 7). Tissue culture is a way of conducting asexual reproduction on a massive scale; the process is now a routine laboratory and commercial procedure, and examples can help students' appreciation of the extent to which this is part of everyday life. A video sequence is useful to trigger debate on the subject. Students should always be encouraged to discuss cloning in a balanced way in order to clarify some of the advantages to society of these techniques. Cloning animals is not as easy as cloning plants.

Cancer
Few young people have a good understanding of what constitutes cancerous growth. You can provide help by making the link with asexual reproduction in terms of uncontrolled cell division (mitosis). A video sequence can demonstrate the irregular mass of cells (tumour) that may have come about as a result of mutation in the genes that control cell division (see Chapter 7).

Useful second-hand data can raise older students' awareness of the way in which scientists begin to correlate the incidence of a disease with a particular factor. This is also an opportunity to talk about different carcinogens (ionising radiation, certain chemicals) and some viruses, all of which are linked with the promotion of cancerous growths in both plants and animals.

Treat the topic with some sensitivity since a student may have/ have had a relative or friend who has suffered from, or is being treated for, cancer.

Further activities

- Spring: record numbers of duckweed in a particular area of a pond over several days to demonstrate the rate of increase.
- Try some common horticultural methods:
  - Late spring – stem cuttings (geranium, *Coleus*); leaf cuttings such as *Begonia rex*; layering ('pinks')
  - Early summer – rooting the 'buds' of runners (strawberry).

- More advanced students could research tissue culture procedures, finding out how a callus is formed and how tiny plants are grown from subdivisions of this. Introduce this activity with a video sequence together with literature from commercial plant breeders (e.g. producers of hybrid orchids).

Enhancement ideas

- Examine yeast cells for signs of budding. Under the microscope students can watch cells reach a certain size and produce outgrowths (buds), which eventually split to form new individuals. The use of an eyepiece graticule on the eyepiece lens of the microscope allows students to judge the size of the cells. A digital microscope linked to a data projector/screen provides for a whole class observation and structured discussion.

How Science Works

- Extend the work for investigative activities in which advanced students might explore the effects of culture temperature, or the amount of sugar, on the rate of cell division. Withdraw a drop of culture from the starter flask every 15 minutes and count the number of cells in the field of view. Plot a line graph to illustrate the increase in numbers of cells over time or pool class data entered on a spreadsheet and obtain mean values and plots of rates of the increases in numbers.
- Using Figure 6.1, ask students to explain what might be happening at X, Y, and Z. Explore possible reasons for why yeast eventually stops budding.

Figure 6.1 The growth curve for yeast

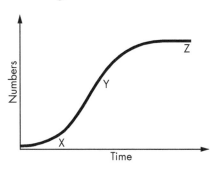

6.2 Characteristics of sexual reproduction in plants and animals

Previous knowledge and experience

Students will have met the idea of life cycles and noted the main stages of the human life cycle: humans produce babies that grow into children and then into adults. Similarly, they will have been introduced to the main stages of the flowering plant including growth, pollination, seed dispersal and the germination of seeds to form new plants.

A teaching sequence

The sequence should try to explore the question, 'Why sexual reproduction?' Broadly it should develop understanding that there is:

- an increase in numbers of individuals
- variation amongst offspring of the same species
- a wide distribution of individuals.

By considering and comparing different reproductive patterns in other animals, students can be helped to forge links with the human reproductive pattern (Section 6.3).

■ Examples of plant sexual reproduction

Use and exploit any opportunities you have to check students' understanding of plant reproduction and different plant life cycles, building on teaching sequences from earlier stages. The idea that many plants can reproduce both asexually and sexually should be reinforced, using examples from across the plant kingdom as well as flowering plants, including grasses.

Many flowering plants have interesting mechanisms for the production of specialised gametes; the male gametes are the pollen cells, found in the anthers; the female gametes are the ovules, found in the ovary. Meiosis is the special cell division that produces these gametes (see Chapter 1). Students can be introduced to the idea that many (but not all) flowering plants are hermaphrodite, with the capacity to produce both male and female gametes. Try to present examples of single sex plants, such as the separate male and female holly plants (genus, *Ilex*). Students could compare these with other types of plants (e.g. birch) which are hermaphrodite, producing both male and female flowers on the same plant. Students could then be asked to suggest mechanisms which prevent self-fertilisation and explore the advantages of out-breeding.

145

Further activities

- Soaked broad bean seeds provide younger students with independent, observational homework (over 4–5 weeks). Each student needs a jam jar and a roll of thick, absorbent paper, ensuring the seed is clamped to the side of the jar. The paper sits in 2 cm of water and acts as a wick. Ask your students to keep written diaries, supported by drawings, for the whole life cycle.

Enhancement activities

- Students might consider how fruits are formed without seeds (e.g. seedless grape, satsuma). (These are generated by spraying with hormones that stimulate fruit production.) They can deduce that the crucial stage of fertilisation has been omitted and the resulting fruit is of no use in producing another generation.
- With more advanced students, illustrate the starch food store in seeds and demonstrate the role of the enzyme amylase, using halved, soaked barley seeds placed with the outside in contact with starch agar in a Petri dish. After 24–48 hours at room temperature, flood the dish with iodine solution. The zone around each grain remains clear while the iodine on the rest of the starch agar is blackened. Ask students to account for the lack of starch in the clear zone.

Eye protection is needed when handling iodine solution.

How Science Works

- Interesting investigations with small seeds and fruits can be carried out. How do particular seeds fall? Ask students to use dandelion or sycamore seeds in order to examine the relationship between height above ground and rate of falling. Ask them which variables (mass, size, shape, etc.) might be changed in order to slow down (or speed up) the rate of fall. Data can be entered into a spreadsheet for further graphical analysis. Model seeds can be made to test ideas further. Ask them why a slower rate may be an advantage to the plant producing the seeds.

■ Examples of animal sexual reproduction

The idea that most animals reproduce sexually as a result of the production of specialised gametes needs reinforcing for younger students; the male gametes are spermatozoa (singular: spermatozoon), abbreviated to sperm; the female gametes are ova (singular: ovum). Meiosis is the special cell division that produces the gametes (see Chapter 1).

Students can be helped to think about the reasons for sexual reproduction, particularly mammalian reproduction, which is technically more complex than asexual reproduction. Encourage them to deduce that it requires more energy and specialised organs that produce gametes, and also results in fewer offspring. The crucial advantage is that individuals are slightly different from either parent and from other offspring of the same parents. In a changing environment some may be well-suited and are able to exploit new resources or colonise new areas. An organism's 'fitness' for a particular environment is a useful concept since it links environmental features and the organism's individual characteristics (see Chapter 8).

■ Specialised sex cells; external or internal fertilisation; courtship

Students should begin to appreciate that almost all organisms can reproduce sexually, even those that also reproduce asexually. Examples might include life in a coral reef, including invertebrates such as sea anemones, corals and jellyfish (e.g. see filmed sequences from *Malice in Wonderland* (1994), available from BBC Active Video).

Such sequences present visually the specialised sex cells (gametes), one from each parent of the same species, which join to form the first new cell (zygote) of the new individual. Examples should also show that there are some individuals able to produce both male and female gametes; these are described as hermaphrodites (e.g. most flowering plants, earthworms and many molluscs such as snails and slugs).

By providing stimulating materials and sources of information including texts, CD-ROMs, pictures and video clips, your students can be encouraged to study the life cycles of certain animals in more detail (e.g. fish, frog):

- Discuss how eggs are fertilised externally/internally, pointing out the numbers of eggs that are fertilised at any one time. Students can be helped to think why this is so.
- Ask students to identify any pattern in the number of eggs/ internal and external fertilisation/internal and external development/aftercare by one or more parent/growth pattern/ chance of offspring surviving to maturity.

Video sequences can be well deployed to illustrate courtship patterns, as well as the act of copulation and fertilisation, in a variety of animals, such as amphibians, fish and birds. BBC *Trials of Life* (1990) has two episodes, *Courting* and *Continuing the Line*, which illustrate a variety of breeding activity (BBC Active Video).

■ External versus internal development; aftercare

Discuss the advantages of retaining the young offspring in the body of a parent (usually the female – but note, for example, the use of the male pouch by seahorses). You can draw on students' own experiences of a variety of newborn pets, and on video clips of a variety of mammals' offspring immediately after birth. Pose the question, 'Which types of mammals are more dependent on having continuous access to nutrients and to protection from harm, including predators?'

Further activities

- Students can discuss the extent to which humans and other mammals provide for emotional as well as physical needs of their offspring in their aftercare.

Enhancement activities

- The advantages of sexual reproduction may be seen also as the disadvantage of asexual reproduction, where only a stable environment will ensure reproductive success. Students can compare and then summarise in a table the advantages and disadvantages of both types of reproduction.

6.3 Human reproduction and sex education

■ Legal framework for Sex and Relationship Education (SRE)

You are likely to be involved in teaching aspects of sex education whether or not you are a biology teacher though the extent to which this is the case varies from school to school and between countries. In England, governors of maintained secondary schools and special schools with secondary-aged students are required by law (Education Act, 1996; Learning and Skills Act, 2000) to provide sex education and relationship education (SRE) for all registered students, including those over compulsory school age. SRE has been defined as 'learning about sex, sexuality, emotions, relationships, sexual health and ourselves' (Sex Education Forum, 2005).

However, the current law does not yet define the purpose and content of sex education other than it *must* include education about HIV and AIDS and other sexually transmitted infections. All such schools must have a policy on sex education with copies available for parents and for Ofsted inspection.

Ministerial guidance for SRE (DfEE, 2000) has emphasised developing knowledge, skills and attitudes and appropriate teaching methods. It recommends that puberty, menstruation, contraception, abortion, safer sex, HIV/AIDS and STIs should all be covered within a planned programme as part of Personal Social and Health Education (PSHE) and Citizenship. Parents have had the right to withdraw their child from all or part of SRE provided that this is *outside* the requirements of the science National Curriculum.

Where your school has a well-formulated and clearly expressed sex education policy, the ground rules and boundaries for any teacher of sex education will be in place. You should also be supported in knowing the lines of referral within your school if a particular Child Protection situation arises. Assisted by the Children and Young Person Act, 2008, ministerial and Ofsted guidance now asks schools to ensure:

- the school has an effective child protection policy, known to staff and parents
- a senior member of the teaching staff acts as the 'designated teacher'
- the procedures of the Area Child Protection Committee are followed by school staff where there are concerns about students (sharing concerns, referrals, written records, attendance at meetings, monitoring children on the child protection register etc.)

You should always provide invited speakers and all other visitors with a copy of the school sex education policy and ask them to abide by it.

Your own classroom relationships with students are important for being effective in this sensitive area, and you should not be deterred from encouraging, wherever possible, open and frank discussion about issues, though never about particular personal circumstances. Where a student might approach you to share some personal information, you should use your professional judgement about keeping confidentiality, while at the same time *never promising* confidentiality. Remember that if you provide details of alternative sources of advice and of treatment this does not count as sex education, so always be guided by the school policy.

Some areas of the broader sex education curriculum, particularly how to approach sexual orientation and the issue of children's rights, can remain rather uncertain for some schools. Sex education embraces both the reproductive process and sexuality, and therefore benefits from a positive acknowledgement of all types of relationships including homosexual and bisexual ones. For a more detailed discussion of these aspects, see Harrison, 2000, pp. 120–123; Martinez, 2005b.

Developing your own sensitivities to the personal circumstances of individual students and their families is important in your class work, and you should also be aware of potential signs of child abuse. Identification is often not easy and, in addition to more obvious physical signs such as bruises, burns, bites and scars, a general indicator is often neglect. Indicators of sexual abuse include sexually transmitted infections, recurrent urinary infections, sexually explicit behaviour, young students with a lot of sexual knowledge, sexually abusive behaviour towards other children and unexplained pregnancy. Emotional abuse is often indicated by low self-esteem, lethargy or attention-seeking behaviour, and delayed social development. A Child Protection Register is a record of all children within a local authority; an area child protection committee has full details of the Register. As a teacher, always take steps to share any concerns with the designated member of the school staff, writing down details as well as reporting them orally.

Previous knowledge and experience

Students may have been given considerable information and understanding about aspects of human reproduction and relationships, but research shows this delivery can be patchy. Students' physical and emotional developments are highly variable and will partly determine the extent to which everyone has absorbed or questioned information given earlier on. There will be differences in the provision from different homes, social and cultural backgrounds, different primary schools and different teachers within any school.

National surveys to find out what young people really want in this area generally find that much information is often provided too late (UK Youth Parliament, 2007). Targeting younger secondary students is therefore important. Boys generally have lower levels of knowledge of reproduction, contraception and contraceptive services than do girls, and so a school needs to take steps to ensure that boys are not excluded from parts of its sex education curriculum. Girls have indicated that they want more adequate discussion and explanations in order to counterbalance an overemphasis on biological facts. Clearly boys require this too.

Educational videos in this area of the curriculum should be chosen with care for their appropriateness and always be used in such a way that there is time for reflection, clarification and discussion of sexual issues in the classroom.

A teaching sequence

For younger secondary students you are aiming to extend ideas about human reproduction and to relate the ways the body changes in adolescence to your students' developing knowledge about human reproduction, growth and the menstrual cycle.

Use key terminology with care. Technically, fertilisation refers to the fusion of the nuclei; the fertilised ovum develops into a zygote to form an embryo and subsequently the fetus. Students need to gain an idea of size and number of gametes at the time of fertilisation. More advanced students can summarise the similarities and differences between the structure of an ovum and a sperm. Try to assess and then extend your students' present knowledge of puberty, anatomy, conception and its prevention, the development of relationships and the medical problems associated with early sexual involvement.

An important teaching aim is to allow young people to rehearse the forms of language (possibly both slang and formal words) and, through this, to get used to the feelings they have about the language with which they will need to communicate effectively with their partners and health professionals later on.

Bear in mind that some students may have the technical vocabulary but their actual understanding can be poor. They need help in knowing where various anatomical parts are located as well as what the parts actually do. Aim to make your audience feel comfortable about not knowing things, while at the same time making it possible for them to find out. Laughter and appropriate joking can help to make everyone feel good while at the same time enabling you to address every question seriously and with respect.

As a science teacher you are contributing to two spiral curricula in this topic: one is within the statutory science National Curriculum; the other is within the school's (non-statutory) programme for PSHE. In the latter there is a national framework for PSHE and PLTS (Personal Learning and Thinking Skills). One of its cross-curricular dimensions is Healthy Lifestyles. Thus, in teaching about human reproduction in science, you may find yourself planning to develop key concepts through concrete examples that can be applied to real-life situations. For example, you may be helping them *assess risk* in situations that affect aspects of their health and well-being. As part of the overall sex education programme, students will be learning about human reproduction, contraception, pregnancy, sexually transmitted infections, including HIV, and the physical and emotional changes of puberty. Consequently you may find yourself teaching in science about the facts and the laws and the personal and social consequences around drug, alcohol and tobacco use and misuse, and helping students to assess risk-related behaviours during pregnancy.

■ Physical and emotional changes that take place at puberty and during adolescence

Between the ages of 10 and 14, most young people will be entering puberty and will be interested in hormones, how they will be affected by them, the menstrual cycle, wet dreams, erections, fertility, pregnancy (including how it can be avoided) and safer sex. They may also be wondering if their physical development is 'normal'. They will want to know about the difference between sexual attraction and love and whether it is normal to be attracted or in love with someone of the same sex. Young people will be asking questions about relationships, when is the right time to have sex, how to avoid pressure and where they can get more information if they need it, including the best websites, confidential services etc. The FPA has helpful leaflets: *4 Boys* and *4 Girls,* aimed at students aged 12+, about growing up, sex and relationships.

An exploration with your students of how males and females differ physically should enable you to summarise the key changes at puberty. A useful ice-breaking activity might be to place each key point describing a secondary sexual characteristic (e.g. breasts, wider hips, facial and body hair, voice changes, stronger body smell) on a separate card; small groups of students can then discuss and arrange the cards under one of the two headings 'males' and 'females'. Brook has a useful range of classroom leaflets for younger students available in packs of 50: *Looking Ahead – Girls* and *Looking Ahead – Boys.* A further useful resource can be found at the Channel 4 Learning Shop, where there is a CD-ROM *Human Body: In Control* (for 11–16 years) covering puberty and the menstrual cycle. You can find details at www.4learningshop.co.uk.

Stress the wide variation in the age of onset of puberty and the, generally, earlier age of onset for girls. Help them understand that changes in hormone concentrations result in the development of secondary sexual characteristics and emotional changes at puberty. Explore possible reasons for the earlier onset over the last 50 years. Although early onset may have a genetic component, the optimum height/weight ratio is also related to improved living conditions and healthier lifestyles, including diet. Puberty also extends over some years for an individual, as illustrated in girls by the time taken for a consistent menstrual cycle.

How Science Works

- Ask your students to investigate the range of heights amongst males and females within a particular age group (or age groups) for a particular population. This is best done using a spreadsheet for all data, to include height along with many other variables such as eye colour, tongue rolling and handedness. Try to avoid investigating 'body weight'. Adolescents can be particularly sensitive about their body image and the position they hold in the peer group. If poorly handled, any of these activities can lower a young person's self-esteem. Carefully handled, they can allow students to take account of their uniqueness with respect to all the characteristics under scrutiny.

■ Male and female reproductive systems: structures and functions

Many younger students tend to have muddled ideas about how many urinary, genital and defaecatory orifices people have (girls have three, boys have two) and often both boys and girls do not know from where a girl urinates.

The *BBC Key Stage 3 Bitesize* website (www.bbc.co.uk/schools/ks3bitesize/science/organisms_behaviour_health/reproduction/activity.shtml.) is designed to support students in their revision of topics. There are excellent video clips within *The Human Reproductive System*, and the ones on puberty, the male and female reproduction systems, the menstrual cycle, fertilisation and development could be selected and used separately to support your teaching. They provide clear diagrammatic overviews and have an element of humour too.

Key vocabulary for the male genital organs includes: scrotum, testis, epididymis (sperm-producing tube), sperm duct, seminal vesicle, prostate gland, urethra, penis, erectile tissue, foreskin; and for the closely associated urinary organs – kidney, ureter, bladder, urethra; and the defaecatory organs – rectum and anus.

Key vocabulary for the female genital organs is: ovary, Fallopian tube (egg tube, oviduct), uterus (womb), cervix, vagina, clitoris and vulva.

Students should understand that the urethra has a dual function in the male. The similarity in sound and spelling of some terms such as 'ureter' and 'urethra' can also be confusing for some. In the female the urethra has only one function, which is in connection with the urinary system. The vocabulary for the female urinary and defaecatory systems is the same as for the male.

Older students need to know the structure and function of all these parts; younger students can be presented with a reduced list of key terms or simplified terminology. It is generally best to provide accurate, but unlabelled, line diagrams that students can then label themselves. This activity can be extended into a card-sorting/matching exercise in which students are provided with key names on one set of cards and a second set of cards with the key functions of the parts.

Help students to understand that the erectile tissue of the penis becomes firm as it fills with blood when the penis is stimulated either manually or indirectly through specific visual or other stimuli. Some male students may need your reassurance about the normality of wet dreams and masturbation. The dual role of the urethra in the male will need some clarification: glands at the base of the bladder produce secretions that wash away the urine in the urethra since urine can destroy sperm.

Masturbation for female students also needs clarification, particularly as it is rarely discussed openly in school sex education or in school textbooks. Indeed, the clitoris is often rendered invisible by being absent from diagrams as well as absent in any discussion about structure and function (see Reiss, 1998, for further discussion of sexuality in science textbooks).

Muscle rings in the sperm duct squeeze the sperm along the passage. This action can be simulated by pushing toothpaste along in its tube, which is similar to peristalsis in the intestinal tract (see Chapter 2). Further glands mix nutrient secretions with the sperm to form semen. Ask students why this is necessary. Illustrate the volume of the ejaculate: about one teaspoon of semen is produced at ejaculation. The prostate gland is often incorrectly referred to as 'prostrate'! It is frequently enlarged in older men and students can deduce the effect of any enlargement on the frequency of and difficulty in urination.

Since the female reproductive organs are largely invisible, and therefore particularly mysterious, ask students to site the position of the ovaries by placing their fingers on their own abdomen. To do this, suggest they feel for the front points of the pelvis and move in towards the navel an inch or so. Ask students to feel the tip of their nose with a forefinger. Say that this feels rather like the cervix, the ring of muscle that closes the lower end of the uterus where it joins the vagina. Use a model of the human torso, and a skeleton, to help pupils understand the 3D arrangements of the key reproductive organs – most diagrams are presented as 2D arrangements, which can be confusing.

There is a very small hole in the cervix to permit sperm to enter; the cervix dilates during labour; check that students understand why. Students can estimate the size of the adult vagina and the size of the adult, non-pregnant uterus. For comparison use a medium-sized inverted pear (about 10 cm long) and tilt it backwards slightly to illustrate the angle of the uterus with respect to the vagina.

Point out the need for a good blood supply to the uterus and explore why this is so. Ask students to suggest how long the egg takes to be moved from the ovary to the uterus (24–48 hours). The vagina is a muscular tube with sensitive nerve endings and glands that can secrete mucus. Explore with your students the reasons for these.

Sperm swim towards the oviducts aided by movements of the female reproductive system. After an hour they no longer swim but they can survive in the uterus or oviducts for 3 or more days. Explore with your students why it is important to be aware of this and how the timing of ovulation can influence the chance of fertilisation (see also page 159).

Most confusing for younger students are the, often dramatic, video shots of the moment of ovulation, or the cartoon pictorial accounts of the subsequent movements of the ovum in the oviduct. Neither gives any idea of scale or the relationship of the microscopic structures to the macroscopic organs and their actual positions in the body. Always take steps to use the pause button on the video sequence and check your audience fully understands the events portrayed.

■ Gamete production; sexual intercourse and fertilisation

The school sex education policy and the PSHE Lead teacher in the school should guide you in whether you can be open about exploring a range of issues. Beyond the biological content that you are expected to teach, you will also find that typical concerns of young people can arise unexpectedly or might be planned for in your teaching. Some examples of students' questions follow:

- What should I do if I feel I am being pressured into having sex? Is everybody doing it?
- My religion says that being gay or having sex before is marriage is wrong, what should I think?
- In my community being a teenage parent is acceptable – is this wrong?

Further, you may wish, or be required, to take on important safety aspects of the PSHE *Healthy Lifestyles* agenda in relation to sexual health. Some students will have many concerns (examples are listed below), while others will need to be made aware of the particular responsibilities that each partner has in any sexual relationship:

- What is safer sex?
- Should everyone who is sexually active carry condoms?
- What infections can be caught from having sex? What are the symptoms?
- What are HIV and AIDS and how do you get them? Is it always through sex?
- Does sex always lead to pregnancy? How can conception be prevented?
- Are there ways of enjoying sex that don't risk pregnancy or infection?
- What are the different methods of contraception? Are some easier to use than others?
- When should emergency contraception be used?
- Who should be responsible for contraception/safer sex in a relationship?
- If someone is on the pill, why should they use a condom as well?
- Does drinking alcohol or using drugs affect my decisions about behaviour?

This work in the classroom may include the need to provide information (not advice) on where young people can go for specific professional help and advice about sexual health. The following concerns of young people can assist you in planning for the type of information that you might be required to provide:

- If I think I have a sexually transmitted infection, where can I get it treated?
- If a woman gets pregnant, what choices does she have?
- What are the best websites on sex and relationships for young people?
- How can I find out about local contraception and sexual health services, and what should I expect from them?
- Can I see a nurse or doctor in private?

Most seriously, there is little to be said in any textbooks and many teaching materials about legal issues and most particularly about the fact that, in Britain, it is illegal to have sexual intercourse with anyone under the age of 16. Brook has a useful resource for professionals: *Under 16s – The Law and Public Policy on Sex, Contraception and Abortion in the UK* (2006) which can be ordered on their website:

 http://www.brook.org.uk/professionals/application/shop.
From the age of puberty, meiosis in each testis (see Chapter 1) produces immature sperm and these develop into mature sperm in the testes (singular: testis). Statistics are helpful in bringing about a better understanding of male fertility including numbers, size and

motility of sperm. A human sperm is about 1.4×10^{-4} mm long and can travel the 15 cm from the vagina to the neck of the cervix in about 3 hours. Younger students often appreciate analogies such as the challenge for sperm reaching the egg being similar to a human trying to swim the Channel in treacle. Sperm production is highest in the early twenties. The relatively large size of the ovum compared with the sperm can be demonstrated with scaled-up paper models.

Meiosis in the ovaries produces the immature eggs, which are already present at birth. From puberty the female sex hormones control a cycle that produces one mature egg about once a month. Not all students recognise that the two ovaries alternate in this production, so that each ovary only produces an egg about once every 8 weeks. The egg develops in a follicle and one of the important jobs of this follicle is to prepare the uterus for pregnancy. Note that young people might be confused if you say the egg 'moves' from the ovary to the uterus; the egg is actually propelled indirectly by the action of cilia and mucus, from the oviduct to the uterus.

Use video sequences and other pictures to illustrate and discuss the processes, including sexual intercourse, which lead to fertilisation and implantation of the zygote. Students can devise a flow diagram using ICT to represent key events leading to fertilisation; alternatively, ask them to produce a pictorial flowchart as a poster to show how the male and female gametes are brought together. The emotional as well as the physical aspects of courtship and sexual intercourse should be addressed through discussion, though do check that the school sex education policy allows for this.

Textbooks and most teaching materials assume intercourse is heterosexual and that sexual activity is penetrative (vaginal) intercourse. Such texts might also be described as sexist since many words describe intercourse from the male rather than female point of view. Ejaculation by the male is frequently mentioned; orgasm in the female is mentioned only rarely. Diagrams almost always show the so-called 'missionary position' with the female underneath the male. Any accounts of the loving relationships, or indeed the passions or pleasures associated with sex, are notable by their absence. It is therefore important to create a balance between an anatomical account and a psychological and emotional account.

 The website *Teachers TV* contains a video sequence entitled *Bodyscapes – Kiss*, which focuses on the biological and social role of the lips in relation to sexual arousal that could be used to trigger discussion with older students. It can be found at: www.teachers.tv/videos/bodyscapes-kiss.

Your students' feelings, attitudes and experiences will play a part in how they approach the content of this part of the topic. Aimed at parents and teachers, the *BBC Bare Facts* website has some useful lesson plans particularly to assist in getting young people to talk in the classroom: www.bbc.co.uk/schools/teachers/barefacts.

Finally, you can take the opportunity to review fertilisation in terms of fusion of male and female nuclei and discuss how this results in characteristics being passed from parents to offspring (see Chapter 7). Twin and multiple pregnancies are always of interest to young people: provide some statements containing some correct and incorrect explanations and ask them to select the correct explanations of how identical and non-identical twins can arise. This is also an opportunity to establish that, generally, humans have one offspring at a time, and that the reproductive system has developed to make sure that the one offspring is as likely as possible to survive.

Enhancement activities

- Encourage students to speculate about different causes of infertility in men and women, linking their ideas to the structures and functions of the reproductive systems, e.g. low sperm counts, blocked oviducts, infrequent ovulation. This should include information on chlamidyia, cervical cancer (see Chapter 10) and testicular cancer. They could be encouraged to find out more about the technological solutions available, and some of the social and ethical issues surrounding the provision of fertility clinics.

■ Menstrual cycle and the control of fertility; contraception

Menstruation tends to remain a taboo subject in our society, which is unhelpful to an adolescent girl for whom it is acutely realistic. The physical, emotional and practical aspects of 'periods', particularly in the school setting, do very little to reassure girls of the positive experiences of becoming a woman. When young people are asked 'What is menstruation?', research has shown that over one-third of 13 to 14-year-old students do not mention menstrual fluid, and when questioned specifically, the actual source of the menstrual blood is frequently misunderstood. Try to aim for a balance between a purely physiological approach and a more personal account that goes some way to acknowledging the reality of this event for half the school population!

Younger students do not need to know about oestrogen and progesterone changes in the bloodstream, nor that egg development is stimulated by a hormone produced by the brain (follicle stimulating hormone), nor that changes to the uterine lining are controlled by hormones produced by the ovaries. However, try to lay the foundation for later work. Help students construct a diagram of the days in the cycle. Using a 28-day calendar you can ask them to predict when a young woman might expect:

- the day of ovulation
- the day of the start of the menstrual period
- the days when she might be able to conceive.

It is important for students to appreciate that the time interval between day one of the menstrual period and the time of ovulation is variable, both from one individual to another and from one month to another for many woman. Fourteen days after the release of the egg from the follicle, the uterine wall breaks down (enable your students to appreciate that this time interval between ovulation and the start of the menstrual period is a *fixed* time of 14 days) (see also Chapter 5). Students can also mark on their diagram when the uterine lining is thickening, and you can explain that the menstrual cycle prepares the uterus for a fertilised egg in every cycle.

The teaching and associated discussion of the natural and artificial methods of preventing pregnancy (i.e. birth control or 'family planning') frequently falls to the science teacher. Much of what follows is for older students (14+) so be guided by your school sex education policy and mindful that the use of artificial contraception is not acceptable to all adults or young people. The health risks, both physical and emotional, of under-age sex should always be explored fully with your audience. Students need both general and local guidance about how to seek information and advice (e.g. 'drop-in sessions' to a school nurse or youth clinic) and how to buy items or to access the particular medical services that can supply them (the diaphragm, intrauterine device, contraceptive pill and the 'morning after' pill, usually referred to as 'emergency contraception'). In relation to fertility, you may wish to point out that an unfertilised egg will not survive for more than 3 days, although sperm can remain alive for a few days longer. Try to challenge any idea that fertilisation cannot take place if male ejaculation takes place outside of the vagina. A valuable Brook leaflet available for students (14+) is *Sex, Contraception, Relationships and Choice*.

Information on its own about a contraceptive method is of little use to older students; if the curriculum indicates that 'contraception' is to be taught, they should also be shown the correct ways a particular item can be used. This applies particularly to the condom, in both male and female forms, the use of spermicides, withdrawal and the rhythm method ('safe period'), which may involve using the calendar or particular detection methods such as a temperature or a hormone detection kit such as *ClearBlue*. The poorly named 'morning after' pill needs particular clarification since it is actually possible to use it up to 72 or even 120 hours after intercourse. Discussion should always include the relative reliability and risk of failure of the item or method, together with its effectiveness in reducing the transmission of sexually transmitted infections (STIs). Older students need to gain some understanding that, at different ages or stages in a relationship, the use of one contraceptive may be more appropriate than another.

A contraceptive pill is a package of hormones that prevents egg release. A fertility pill is a package of hormones that stimulates egg production and release. Ask more advanced students to suggest suitable hormones for a contraceptive pill and for a fertility pill and to explain their choice. Why would doctors find it much easier to determine a dose for a contraceptive pill rather than for a fertility pill?

Ask older students in groups of three to five to make a video sequence, or PowerPoint presentation, advertising a particular contraceptive. They should consider how the contraceptive works, for whom it would be most useful, and present the chief risks. Alternatively, using desk top publishing, small groups of students might produce a leaflet concerned with one type of contraceptive.

■ Sexually transmitted infections (STIs)

A detailed study of sexually transmitted infections (STIs) falls best within the teaching of microbiology (see Chapter 10) and the modes of transmission should be considered for the six commonest STIs: chlamydia (bacterial), genital warts (viral), gonorrhoea (bacterial), genital herpes (viral), hepatitis B (viral) and Human Immunodeficiency Virus – HIV (viral). Other STIs might include syphilis (bacterial), pubic lice (crustacean), urethritis (bacterial), thrush (fungal), bacterial vaginosis and trichomoniasis (protozoon).

Reinforcement activities can encourage students to consider the routes of infection and importantly, since several of these diseases are life threatening, to consider how they are passed on to another person and how they are best avoided. Offer the reassurance that, with early detection, almost all may be successfully treated. Aim to dispel any misunderstandings. Find out what your students do, or do not, know about STIs; challenge prejudices about those who may catch an STI. Students need to know that people infected with HIV do not have to look ill.

Human papilloma virus (HPV), sometimes called 'genital wart virus', can be passed from one person to another through sexual contact. Younger pupils (11+) need to be prepared for the recently introduced HPV vaccination programme. In the UK all girls in Year 8 (aged 12–13) are offered the HPV vaccine in the form of *Cervarix,* involving three injections over 6 months by a nurse. Explain to all students the known links between some types of

HPV and genital warts, changes to the cervix that may develop into cervical cancer, and changes to the vaginal tissue that may develop into vaginal cancer. As a result of girls taking up the vaccination programme, it is expected for example that at least seven out of ten cancers of the cervix may be prevented in the future. The benefits are long-term since it can take 10–20 years for a cancer to develop following HPV infection. This is also the opportunity to talk about cervical screening programmes and how they will be still needed in years to come.

Explore with your students the reason for offering the vaccine to girls from about the age of 12 (they are unlikely to be sexually active and to have caught HPV). Make it clear that the vaccine cannot get rid of HPV if someone is already infected. Completing the course of injections is also important. Introduce the idea of 'risk' and 'risk-related behaviours' to clarify that some, but not all, infections might result in a cancer.

Providing accurate and up-to-date information about sexual health and prevention of STIs must be an important teaching objective. For each STI, make sure students can focus on cause, symptoms, effect on health/complications and treatment. Another excellent Brook publication, *Ask brook about … Sexually Transmitted Infections*, is suitable for use with older students (14+) and could assist with your own planning of lessons. The AVERT website also has a wealth of information, diagrams and advice for people, including teens, about relationships and HIV/AIDS.

Since the routes of transmission involve sexual contact, a second and equally important teaching objective is to raise students' awareness about the need for safer sex. They should be informed about the risks of transmission through the mixing of body fluids, which include semen, vaginal secretions and blood. This is best addressed when teaching about contraception so students can weigh up the 'pros' and 'cons' of the various contraceptives in terms of effectiveness in preventing transmission of STIs.

Teachers and youth workers can feel anxious about their legal position, particularly when it comes to discussing homosexuality with young people. No current legislation prohibits sensitive and sensible discussion about homosexuality and teachers are within the law to educate against prejudice and discrimination. Of course, you do need to work within the framework of your school's sex education policy. There is a lack of information about sexual identity in school textbooks and other teaching materials, and what information there is, often presents homosexuality as a second-best option, particularly for young people, and one that a person might well grow out of. Bisexuality is mentioned even less frequently. In terms of equal opportunities these are serious omissions. Rarely is there mention about the actual expression of homosexual feelings, and in connection with STIs we can see that there are aspects of teaching about safer sex (e.g. during anal intercourse) that might be overlooked.

Finally, older students need information about testing procedures and local GUM (genito-urinary medicine) clinics. Should you be asked by a student about a sexual activity and you do not know what it involves, you or the student can contact the Sexual Health Line (0800 567123) for further information. Young people can also contact NHS Direct (0845 4647) for advice. Inform your students about the high incidences of HIV infection in some parts of the world and promote some discussion about safer sex on holiday. Case study scenarios and 'agony aunt' letters make useful classroom resources for increasing dialogue about the issues.

Schools may also decide to invite health professionals into the classroom to support the teaching in this area, and local SRE Advisors (Local Authority), Teenage Pregnancy Co-ordinators/Managers (NHS) or other personnel from GUM clinics can often provide expert help. There are good video sequences, too, though it is important to check that these are up to date and to plan carefully so young people always have an opportunity to talk about the topics and issues that arise.

Further activities

- Use a three-dimensional model of the human torso with interchangeable male/female/pregnant female parts, together with good-quality three-dimensional drawings or illustrations of the male and female urino-genital tracts. These will allow students to understand the relative positions of key organs and structures. Ask questions about number, shape and size as well as the technical name of the parts. Do not assume that knowing the names means a full and accurate understanding of their functions.
- Let students have the opportunity to look at the male/female reproductive system of a rat that has been dissected to expose these areas. They can identify the key parts and find out what they do or where they lead to.
- Examine the microscopic structure of the testis and ovary. Let students have a look, under the low power of a microscope, at a prepared, stained slide of a mature mammalian testis (i.e. a thin section which has been cut and mounted on the slide). Point out the sperm located in the tubules. The testes produce sperm continuously and store them in the epididymis, typically 350 million sperm in a human, at lower than core body temperature. Ask why it is an advantage for the testes to be situated outside the main body cavity, and whether there may be disadvantages too.

A prepared microscope slide of a mature ovary of a mammal can reveal, under low power, the immature eggs towards the edge of the ovary.

Enhancement activities

- More able/older students can be asked to summarise in a flowchart the way the pituitary and ovarian hormones interact to produce a regular menstrual cycle. Alternatively, provide your class with a circular time chart (Figure 6.2) together with five 'cards' containing relevant information about oestrogen, progesterone, luteinising hormone, follicle-stimulating hormone and the corpus luteum. They should stick the cards on the outside of the chart against the part of the cycle they correspond to, and then number them from 1 to 5 in the sequence in which they occur.

Figure 6.2 Time chart and information cards to make a diagram showing the hormones involved in the menstrual cycle

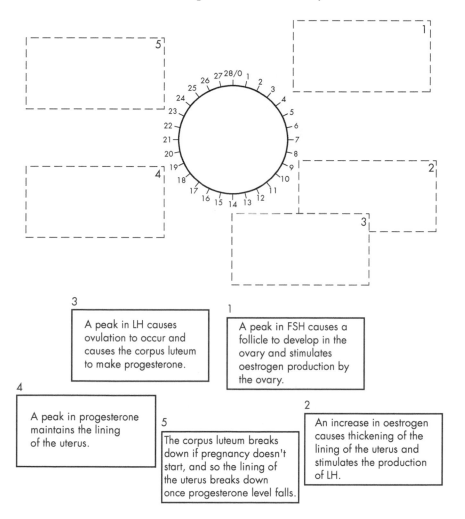

3
A peak in LH causes ovulation to occur and causes the corpus luteum to make progesterone.

1
A peak in FSH causes a follicle to develop in the ovary and stimulates oestrogen production by the ovary.

4
A peak in progesterone maintains the lining of the uterus.

5
The corpus luteum breaks down if pregnancy doesn't start, and so the lining of the uterus breaks down once progesterone level falls.

2
An increase in oestrogen causes thickening of the lining of the uterus and stimulates the production of LH.

Selected parts of *The Male Reproductive Organs* and *The Female Reproductive Organs* from the BBC series *Don't Die Young* (2008) could be used to stimulate discussion with older students. These programmes address related health issues for men and for women (BBC Active Video).

Social and ethical issues

- The control of fertility raises a number of points of social and ethical concern. The following technologies provide opportunity for young people to think about the link between the science and society:

  - Artificial insemination of a woman with her husband's sperm (AIH) allows an infertile couple to have a child without a third party being involved. If the husband cannot produce sperm then sperm from a donor might be used (AID). You can explore with older students how this might be similar to, or different from, adoption of a child by the couple. Ask them to make a list of issues that might be of concern with AID.
  - *In vitro* fertilisation (IVF) (which used to be called making 'test-tube babies') provides a particularly useful discussion topic. It is used to treat women whose ovaries are functioning but whose oviducts are blocked, or where sperm motility of the partner is poor. The woman is treated hormonally to super-ovulate. Ask students, 'What are the risks of IVF?'

6.4 Human pregnancy

Previous knowledge and experience

Younger students are likely to have met some of the processes of fetal development; some understanding of the role of the parents and childcare will also be present.

A teaching sequence

Your aim is to extend students' ideas about how offspring are protected and nurtured following the fertilisation of the egg. Students should learn that the fetus develops within a membranous bag, supported and cushioned by amniotic fluid. They should develop their understanding of the placenta: it supplies nutrients and oxygen to the fetus via the umbilical cord, and removes carbon dioxide and other waste products.

Students will need help in making the links with other important biology topics such as the circulatory system, so they can appreciate the route taken by nutrients from the mother's digestive system to the fetal brain and other tissues. They will need to link their broader knowledge of diffusion gradients to explain how oxygen, water and digested food pass from the mother's blood to the fetal blood in the placenta, and, in the reverse direction, how carbon dioxide and other waste materials leave the fetal blood and enter the maternal blood.

■ Fetal development

Use photographs, models, diagrams, video clips, CD-ROMs or ultrasound scan pictures to look at the immediate and later changes in a developing fetus from implantation to birth. Discuss the sequence with students, using questions to establish their existing knowledge and any misunderstandings.

Explore carefully what happens to the fertilised egg (zygote) in the 6–7-day journey from oviduct to implantation in the uterine wall. Once it is fully implanted, after a further 6–7 days, its cells begin to produce a hormone. It is this hormone that is detected by most pregnancy tests.

Several teaching schemes include photocopiable templates for making models of the fetus in the uterus. Identify the key structures within the pregnant uterus and explain the functions of the amnion and amniotic fluid. Ask students to think about how the fetus may obtain its nutrients and oxygen before the formation of the placenta. Give them a diagram to label and use arrows to show the directional movement of all the substances.

The notion of spontaneous abortion (miscarriage), which is particularly high in the first 2 weeks of gestation (i.e. pre-implantation), might be discussed, together with the continued risk of miscarriage in the first 3 months of pregnancy. Many changes take place in a woman's body during pregnancy and hormones play an important role (see Chapter 5). As the implanted embryo grows, the tissues of the uterine lining and of the embryo itself develop into a placenta. At 12 weeks the placenta is fully formed. The embryo is now referred to as a fetus.

The FPA has leaflets available in comic-book format suitable for students aged 12+; see *Pregnancy – a young person's guide*, which considers the preparation for a pregnancy, the developing fetus and birth. The development of the human embryo during the 40-week period of gestation lends itself well to some descriptive writing by students.

■ The placenta

Ask the students for which body organs the placenta acts as a substitute. Stress that fetal and maternal blood supplies are very close but completely separate. Explore the mechanism by which oxygen, glucose and amino acids can pass from mother to fetus, and by which carbon dioxide and other waste substances, such as urea, can pass from fetus to mother.

A number of health-related issues can be explored to prepare older students as future parents, together with the responsibilities that this entails. Ask which harmful substances or organisms might pass from mother to fetus. Students need help to appreciate that harmful substances can cross the placenta to the fetus and affect development. Challenge more advanced students to explain why, at one time, teenage girls were offered the *rubella* (German measles) vaccine but boys were not. Students can be offered secondary sources to find out more about the passage across the placenta of alcohol (and other drugs), including nicotine from cigarette smoke. They could make a poster or leaflet, which explains to young women how smoking or alcohol can affect a developing fetus from the earliest stage of pregnancy.

■ Birth

The processes of birth can be summarised by students using photographs and diagrams as illustration and alongside a flowchart that shows the main stages of labour. Explain the process of birth as involving the muscles of the cervix relaxing and the uterine muscles contracting, resulting in the fetus being pushed out, usually head first, with the placenta expelled afterwards. Describe how the newborn baby obtains nutrients and oxygen vital for survival and growth.

The many excellent videos and photographs should always be used with care, to avoid embarrassing or distressing those students who may find that some sequences make them feel squeamish or faint. Always be sensitive, where pregnancy and birth are concerned, to their impact on young people and their families (e.g. adoption, miscarriage or neonatal death).

Students should also be helped to understand the importance of breast milk in providing antibodies to protect against infection from common microorganisms. They could find out more about newborn reflexes, such as head turning when a newborn's cheek is touched. Students can also summarise the care that is needed by newborns and older babies (link with Section 6.2).

Further activities

- The whole topic of pregnancy, birth and neonatal care provides good opportunities for students to get to know more about the medical, maternity and welfare services provided by GPs and other parts of the NHS, as well as parenting issues. A school nurse, local midwife or health visitor may be helpful in providing your class with accurate information and detail about Caesarean sections, induction of birth, breech births, modern monitoring techniques, good-quality childcare and so on. In advance of a session with a visitor, let students draw up a list of possible questions they might ask.
- You can help students prepare for talking with professionals about health-related aspects with a range of suggestions (e.g. Why are pregnant women offered additional iron supplements? Why can obesity be a health problem in pregnancy? Do fathers have to watch the birth? What is a good role for a new father?). Students can find out more about the composition of breast milk and the value of the colostrum that is produced immediately after birth as the first feed.

Enhancement activities

- Survival rates of very small and very large birth-weight babies can be low. Students can find out more about these conditions and think about the relevant factors (e.g. the link between smoking and lower weight babies).
- *The First Fourteen Days* (1990) provides valuable pictures, together with more challenging possibilities with specific discussion of human embryo research for older students (BBC Active Video).
- A premature baby aged 28 weeks can be helped to survive. Students can find out what a baby looks like at this age. Ask them what an incubator has to provide for such a baby. *Miracle Baby Grows Up* (BBC Active Video) looks at the religious, moral and social issues of premature babies and whether science, technology and medicine should still be employed to assist those with less chance of survival.

Social and ethical issues

- Abortion is a particularly sensitive issue that needs careful handling by you, and with due consideration for some religious and cultural groups. You should endeavour to remain neutral at all times when presenting the debate and allow your students to work out for themselves the position they may wish to adopt. Be particularly cautious about how you might use the materials and resources produced by some of the key pressure groups. The FPA resource pack *Why Abortion?* is aimed at young people aged 14+.

- Cloning livestock embryos is an important technique in animal breeding (animal clones are genetically identical individuals formed by carefully dividing a 16- or 32-cell stage ball of cells, or by growing a ball of cells from some maternal body cells, as was the case in Dolly the sheep). The ball of cells is planted into the prepared 'pregnant uterus' of the recipient. Ask students to summarise the advantages of such processes in animal breeding.

Equipment notes

■ Asexual reproduction in plants and animals

Bread mould cultures require moistened pieces of stale bread in dishes exposed to the air for 24 hours. Keep the bread slightly moist to prevent it drying out, and cover with a glass jam jar to keep it humid.

Ensure that students do not lift the jar.

A **cell suspension of actively growing yeast cells** can be made at the bottom of a small flask, using 8 g fresh baker's yeast and 10 g glucose (or cane sugar), made up to 200 cm^3 with distilled water. Plug with cotton wool and leave in a warm room (22 °C) for 20 minutes before use with the class. Always pre-test culture conditions before the lesson to ensure that cell division is taking place. In a small pipette, transfer one drop of the culture to a microscope slide with a small amount of methylene blue stain and cover with a coverslip. Use methylene blue for staining living cells as follows: 1 g methylene blue; 0.6 g sodium chloride; 100 cm^3 distilled water.

Methylene blue is harmful.

■ Sexual reproduction in plants

A specially selected mutant **rapid-cycling *Brassica*** provides a versatile and easily maintained resource to develop students' understanding of the life cycle of a flowering plant in 4–6 weeks, when grown under a specially constructed artificial light bank in the school laboratory. Its versatility lies in the opportunity to study its life cycle at any time of the year; it can be 'grown to order' to obtain germinating seedlings, pre-flowering or flowering plants for a particular teaching day. There are numerous investigative ideas available for teachers as part of extensive resource packs. Contact Science and Plants for Schools (SAPS).

Starch agar plates are made up as follows: make a starch suspension with 10 g starch and 1 dm³ distilled water. To do this, mix a little of the starch with cold water; bring the rest of the water to the boil; add the starch mixture to the boiling water. To make up the iodine solution use 3 g iodine crystals and 6 g potassium iodide. Dissolve the potassium iodide in 200 cm³ distilled water, add the iodine crystals and make up to 1 dm³ with distilled water. Make up 24 hours before it is to be used to allow the iodine to dissolve fully.

Iodine crystals are harmful; use gloves and eye protection when handling.

■ Sexual reproduction in animals

Using a scale of 1 : 200, students can make scaled two-dimensional models of an egg and a sperm. To do this for the egg, draw a circle with a diameter of 20 cm. For the sperm draw a head diameter of 0.8 cm, and overall length including head of 10 cm. The actual dimensions are indicated on Figure 6.3.

Figure 6.3 Actual dimensions of human egg and sperm. (Based on *Nuffield Coordinated Biology* (1988), Figure 19.1b, p. 245. Harlow, Longman.)

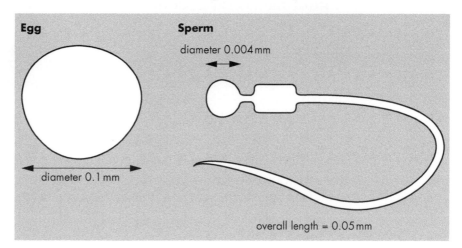

Other resources

Resources and kits

BBC Active Video http://www.bbcactivevideoforlearning.com/.

Family Planning Association, *Contraceptive Display Kit*, to encourage students aged 12+ to think and talk about contraception.

Philip Harris, *Human Torso, Dual sex*, is a life-size model with 28 reproductive parts. Available from www.philipharris.co.uk/.

Philip Harris, *Pregnancy pelvis with mature fetus (40th week)*. Available from www.philipharris.co.uk/.

Additional teaching resources, packs and CD-ROMs

BBC David Attenborough's *The Life Collection*. Now available as 24 DVDs, with beautifully filmed sequences of plant and animal reproduction found throughout. *Life of Birds, Trials of Life, The Life of Mammals* and *The Private Life of Plants* are all relevant to the topic of reproduction. Go to: www.BBCshop.com

Christopher Winter Project (2009). *Teaching SRE with confidence in secondary schools*. CD-ROM resource with spiral curriculum plans, Years 7–11. Go to: www.tcwp.co.uk

Cole, B. (2001). *Hair in Funny Places*. London, Jonathan Cape Ltd. Factual information about puberty in picture-book format, for students aged 9–13.

Harris, R. H. (2005). *Let's talk about sex: growing up, changing bodies, sex and sexual health*. London, Walker Books Ltd. Amusingly illustrated, frank and reassuring information, covering different sexualities, disabilities, ethnicities and body shapes, for younger students aged 10–14.

Harris, R. H. (2004). *Let's talk about where babies come from*. London, Walker Books Ltd. Humorous cartoons with clear information about how babies are made and contraception. Level and language for slightly higher age than the target age of 10+.

Local organisations

Look out for:

- AIDS support services
- Department of Genito-Urinary Medicine (GUM)
- Family Planning Clinic and/or Young Person's Clinic
- Health Promotion Centre
- Lesbian and Gay Communities Resource Centre
- Rape Crisis Centre.

National organisations

AVERT: International HIV and AIDS information and support: www.avert.org/

Brook: Advice and information for under 25s and professionals within the UK on STIs/contraception/relationships/pregnancy: www.brook.org.uk/

Family Planning Association (FPA): UK Library and information service for sex and relationships education and other subjects relating to sexual health: www.fpa.org.uk/

National Aids Trust: UK campaigning charity, also providing information and support on HIV and sexual health: www.nat.org.uk/

Sex Education Forum, c/o National Children's Bureau: Collaborative network, representing over 50 member organisations in the UK; all are involved directly or indirectly in the provision or support of sex and relationships education (SRE). It provides SRE resources, web materials, and training: www.ncb. org.uk/sef/home.aspx

Sheffield Centre for HIV and Sexual Health: National organisation concerned with promoting positive, holistic sexual health and wellbeing and reducing health inequalities. Newsletter, publications and resources for professionals: www. sexualhealthsheffield.nhs.uk

References and recommended books

Blake, S. (2002). *Sex and Relationship Education: A step by step guide for teachers.* London, David Fulton.

Blake, S. and Muttock, S. (2004). *Assessment, Evaluation and Sex and Relationships Education: A practical toolkit for education, health and community settings.* Spotlight series, National Children's Bureau.

DCSF/PSHE Assocation (2010). *PSHE Education Guidance. A summary of government guidance related to PSHE education.* DCSF/ PSHE Association.

Department for Children, Schools and Families (DCSF) (2008). *Government Response to the Report by the Sex and Relationships Education (SRE) Review Steering Group.* London, DCSF.

Department for Education and Employment (DfEE) (2000). *Sex and Relationships Education Guidance, Circular 0116/2000.* London, DfEE.

Harrison, J. K. (2000). *Sex Education in Secondary Schools*. Buckingham, Open University Press.

Martinez, A. (2005). *Effective Learning Methods: Approaches to teaching about sex and relationships within PSHE and Citizenship*. National Children's Bureau. Factsheet highlighting teaching approaches for effective learning about acquisition of information, essential life skills and exploration of values and attitudes. (Also downloadable from the Sex Education Forum website.)

Martinez, A. (2005). *Sexual Orientation, Sexual Identities and Homophobia in Schools*. National Children's Bureau. Factsheet supporting teachers challenging homophobia; suggests ways of addressing diversity and difference. (Also downloadable from the Sex Education Forum website.)

Reiss, M. (1998). 'The representation of human sexuality in some science textbooks for 14-16 year-olds', *Research in Science & Technological Education*, **16**, pp. 137–149.

Sex Education Forum (2005). *Sex and Relationships Education Framework*. London, SEF.

UK Youth Parliament (2007). *Sex and Relationships Education: Are You Getting It?* London, UK Youth Parliament.

7 Genetics and genomics

Jenny Lewis

7.1 Genes
- Who has genes?
- Where are genes found?
- Alleles

7.2 Patterns of inheritance
- Genes and chromosomes
- Chromosomes and sex determination
- Genes, alleles and phenotype
- Single gene (monohybrid) inheritance
- Discrete (single gene) versus continuous (multifactorial) characteristics

7.3 Gene expression
- DNA: the basic structure
- DNA: coding the information
- Reading the message

7.4 The genetic basis of variation
- The role of sexual reproduction and implications for gene frequencies within the gene pool

7.5 Genetic technologies
- DNA fingerprinting
- Genetic screening
- Cloning
- Genetic engineering
- DNA sequencing
- Social and ethical implications

7.6 Genomics
- Cancer development
- Epigenetics
- Stem cells
- Social and ethical implications

Choosing a route

In the aftermath of the Human Genome Project, new developments in genetics, and their possible impact on society, are reported almost daily in the media. As a result, you might expect students to find the topic of genetics fascinating. This is not automatically the case – some students find genetics too difficult and abstract and don't always see the relevance to their own lives.

It is worth spending a few moments thinking about some of the reasons for this as they may influence the way you choose to approach the teaching of this topic:

- Genes cannot be observed directly and are often identified through their effect at the whole organism level, leaving many students with the idea that genes are an abstract, theoretical idea:
  - making explicit the basic structure (a sequence of DNA bases), and the link between a gene and its products (other nucleic acids, polypeptide chains, proteins) may help students' understanding of gene expression and gene technology
  - making explicit the relationship to other biological structures (chromosome, nucleus, cell, etc.) may help students' understanding of the process of inheritance.
- Genetics is complex – gene expression is influenced by the environment (both internal and external) and has an impact at all levels of biological organisation (from cells to whole organisms to populations):
  - making these interactions, at different levels of organisation, explicit may help students to develop a better understanding of gene expression.
- Genetics uses a large vocabulary of specialist terms that may overwhelm and confuse:
  - explaining key ideas in everyday language before introducing the specialist terms may help students to understand the concepts better and so make the scientific terms more meaningful.

The route through this topic is potentially quite flexible but, keeping your students' possible difficulties in mind, it is a good idea to start with a consideration of genes – how they might be defined and how they relate to other biological structures. This will help your students to develop the concept of a gene as a real entity with a structure and a location, and provide them with some basic knowledge to build upon when learning about inheritance or gene expression. Gene expression and inheritance can be taught in either order, and provide a base on which some understanding of variation (the causes and consequences) and gene technology can be developed.

When helping students to develop their understanding of basic genetic concepts, it can be useful to reduce complexity by adopting a traditional, linear view of gene expression (one gene, one protein, one characteristic) but there is a risk that this will result in a deterministic view of genetics in which every characteristic is determined by a single gene. The reality, unexpectedly confirmed by the Human Genome Project, is that there are very few single gene characteristics or disorders in humans. Rather, the relationship between the genome (the entire DNA sequence), gene expression and the environment was shown to be considerably more complex than anticipated. The result is a move away from a focus on single genes (genetics, understood narrowly) and towards a consideration of the whole genome and its interactions with the environment, both internal and external (genomics). Some of the implications of this are presented in Section 7.6 – these ideas could be taught as a final session or could be integrated into the earlier sessions, so you might want to read Section 7.6 before you start teaching this topic.

Within this chapter, genetics is discussed in the context of eukaryotic organisms (animals and plants) in which the DNA is combined with an assortment of other materials including proteins and organised into chromosomes. The basic concepts are the same for all organisms but the organisation of genetic material will be different – for example, bacterial cells have a circle of DNA rather than sets of chromosomes.

7.1 Genes

Previous knowledge and experience

At the most basic level, students will need some concept of living things, including an awareness that organisms are made up of cells and that all plant and animal cells have the same basic structure – a nucleus surrounded by cytoplasm and contained within a membrane. A prior discussion about some aspect of genetics of interest and relevance to your students would give this topic meaning and help to motivate them.

A teaching sequence

As our understanding of genomics develops it becomes increasingly difficult to define a gene in simple terms. For the purpose of teaching genetics at an introductory level the following set of related ideas provides a basic description of a 'gene', which is compatible with our developing knowledge:

- Genes provide information that living things need in order to function; it follows from this that genetic information is found in all living things.
- A gene (usually) has a specific location on a chromosome.
- A gene is (usually) a segment of DNA.
- The sequence of bases within that segment provides information that specifies the possible gene products.*
- Genes need to be activated (switched on) before this information can be translated into a product.
- For any gene, a number of different versions (alleles) may exist.
- Each version contains a slightly different sequence of bases (information) and so results in slightly different products.

These ideas may seem very basic and you might be tempted to embed them in your teaching of inheritance or gene expression, on the assumption that they are self evident and unproblematic – but this is not the case. Research on students' understanding of these ideas shows that even *after* they have been taught genetics, many students seem unaware that all living things contain genetic information, are confused about the relationship between genes and chromosomes (that genes are specific regions of the DNA coiled within chromosomes) and make no distinction between gene and allele. Very few students seem to recognise that there are products associated with a gene. Research has also shown that if these basic ideas are made explicit to students they find it easier to understand the more complex ideas such as inheritance and gene expression, so a little time spent making these basic ideas explicit at the start will save time and enhance understanding later.

■ Who has genes?

To explore and develop your students' understanding of genes and chromosomes you can use a diagnostic question which asks them about some specific organisms. An example is given in Worksheet 7.1. The organisms used in this example were selected for diversity and represent a vertebrate and an invertebrate animal, a flowering and a non-flowering plant, a fungus, a bacterium and a virus.

---

*To teach the basic principles of genetics it is easiest to assume that this product is a protein, but be aware that there may be more than one product and that not all products are proteins (DNA, RNAs of different types and polypeptides are all possibilities).

Worksheet 7.1 Probing students' ideas on genes and chromosomes. In each box answer 'yes', 'no' or 'don't know'

Organism	Do these organisms contain genes?	Do these organisms contain chromosomes?
Rabbits		
Mushrooms		
Bacteria		
Spiders		
Apple trees		
Viruses		

Using this as a small group activity (one sheet per group; aim for consensus) will enable students to exchange ideas and encourage them to justify their views. You will gain most from this activity if you use the feedback session to probe your students' reasoning. For example:

- On what basis do they classify some organisms as containing genes and others as not containing genes?
- What was it that made them unsure about some organisms?
- What are their reasons for thinking that some organisms contain chromosomes but not genes (or vice versa)?

This type of questioning will enable you to identify common alternative conceptions within the class, which you can then address during your teaching. Remember that the point of this exercise is to probe your students' initial understandings about genes. You are not expecting your students to know that in plants, animals and fungi genes are organised into chromosomes, in bacteria they are strung together in a circle of DNA and in viruses the organising structure is variable.

■ Where are genes found?

Most students find it difficult to conceptualise a gene as a real entity with a precise location. This may be one of the reasons why they are confused about the relationship between genes and chromosomes. Talking your students through a diagram that shows the location of genes within the whole organism might help (see Figure 7.1). A number of points could be made explicit:

- All plants and animals are made up of cells.
- All plant and animal cells contain a nucleus (exceptions: mature red blood cells and sieve tubes).
- Chromosomes are contained within the nucleus (exception: during cell division).
- Chromosomes are made up of tightly coiled DNA (and some other materials, including chromatin).
- A gene is a segment of DNA with a specific function.
- Information carried in the gene specifies the gene products.

■ Alleles

Many students have difficulty in understanding the concept of alleles. Such students don't appear to distinguish between a gene (a length of DNA responsible for a particular product) and an allele (the actual information within that gene) and find it difficult to understand what is meant by phrases such as 'alternative forms of a gene'. This difficulty has important implications for their understanding of inheritance. A simple illustration of the difference between a gene and the genetic information within that gene might be useful here. In some breeds of rabbit the fur can be either black or white. That is, there is a single gene for fur colour and there are two versions (alleles) of this gene. One version provides information that specifies black fur; the alternative version provides information that specifies white fur.

Figure 7.1 The relationship between structures

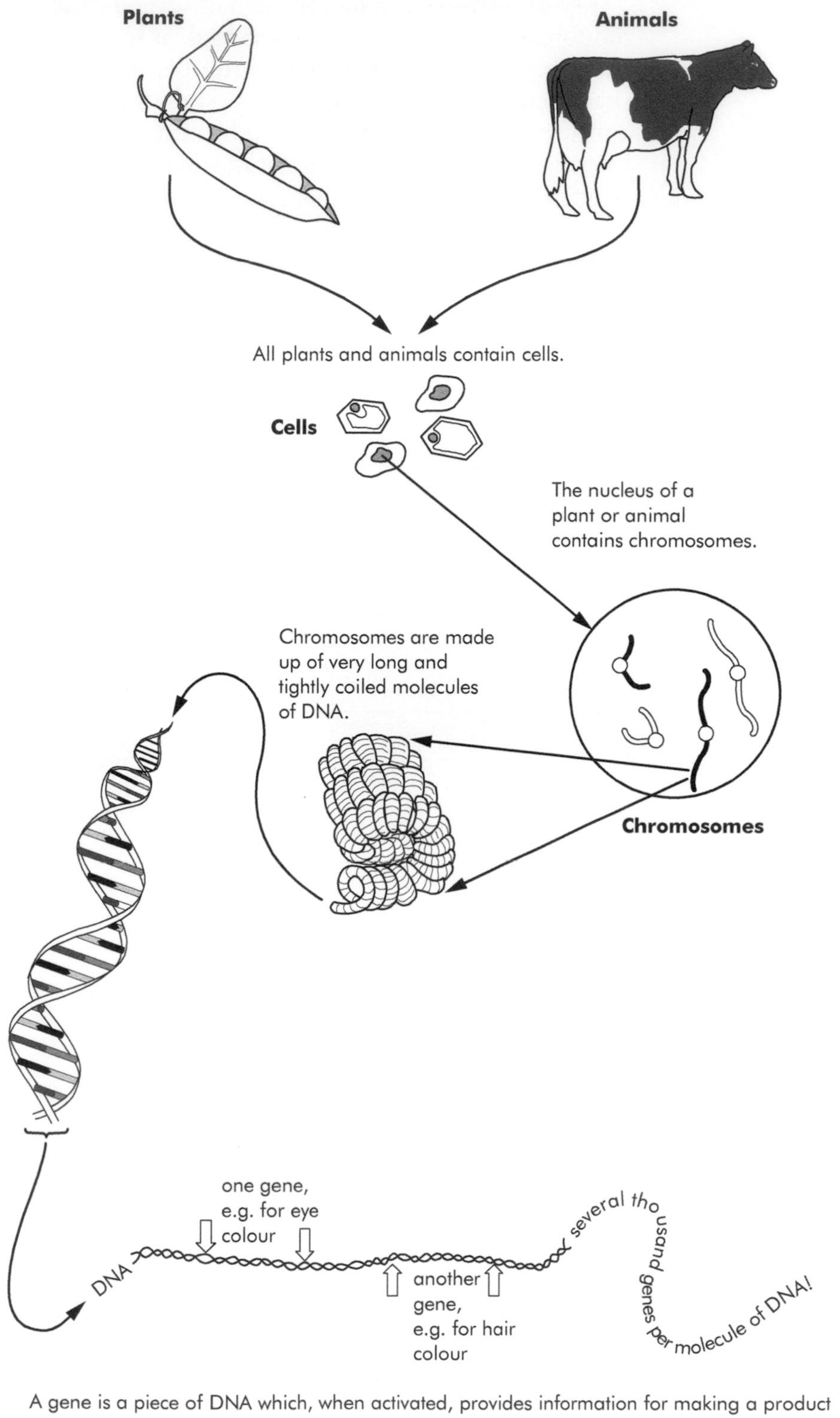

Plants

Animals

All plants and animals contain cells.

Cells

The nucleus of a plant or animal contains chromosomes.

Chromosomes are made up of very long and tightly coiled molecules of DNA.

Chromosomes

one gene, e.g. for eye colour

DNA

another gene, e.g. for hair colour

several thousand genes per molecule of DNA!

A gene is a piece of DNA which, when activated, provides information for making a product

The idea that one gene can exist in several different forms can be reinforced by asking your class to look at some of their own characteristics. Collect data on a number of different discrete characteristics for the whole class – these might include hair colour, hair texture, eye colour and ear lobe attachment but you will need to avoid continuous characteristics such as height and weight. Ask your students to assume that each characteristic is determined by just *one* gene and to draw up a table of results listing the genes they have looked at and the different version of each that they have found. An example of a results table is given in Table 7.1.

Table 7.1 Example of a student's results sheet for collecting data on class characteristics

Characteristic	Variation	Number of students (total = 30)	
hair colour	black	ꟷꟷꟷ ꟷꟷ	7
	dark brown	ꟷꟷꟷ ꟷꟷꟷ ꟷ	11
	golden brown	ꟷꟷꟷꟷ	4
	blond	ꟷꟷꟷ ꟷ	6
	red	ꟷꟷ	2
eye colour	brown	ꟷꟷꟷ ꟷꟷꟷ ꟷꟷꟷꟷ	14
	blue	ꟷꟷꟷ ꟷꟷꟷ	8
	green	ꟷꟷꟷꟷ	4

7.2 Patterns of inheritance

Previous knowledge and experience

Explanations of inheritance usually assume knowledge of a number of different concepts. These include:

- the basic ideas outlined in Section 7.1
- the behaviour of chromosomes during cell division – both mitosis and meiosis
- the idea that chromosomes occur in sets
- the notion that cells may be haploid (contain one set of chromosomes) or diploid (contain two matching sets of chromosomes).

An awareness that gametes (sex cells) are haploid while somatic (body) cells are diploid and an understanding that each gamete contains a unique combination of genetic information would be helpful but cannot be assumed. A substantial minority of students up to the age of 16 do not distinguish between somatic cells and gametes.

A teaching sequence

The term 'inheritance' refers to the transfer of genetic information between generations and the interpretation of that information within the new generation, as reflected in its characteristics (phenotypes). This transfer of genetic information might be between whole organisms (i.e. from parents to offspring during meiosis and fertilisation) or between generations of cells (i.e. from original to new cells during mitotic cell division) – for more details on the process of cell division see Chapter 1.

■ Genes and chromosomes

When students are considering inheritance, the concept of a gene (Section 7.1) needs to be extended to include the idea that genes can be accurately copied, allowing information to be passed on to new cells and new individuals. In addition students need to be aware that:

- chromosomes occur in sets
- within one set of chromosomes each gene (usually) occurs once, at a specific location
- at fertilisation, one set of chromosomes is donated by the female and a matching set (possible exception: a Y chromosome) is donated by the male
- in the new individual each chromosome is paired and so each gene occurs twice (possible exception: genes occurring on the X chromosome)
- each pair of genes may occur as identical or different versions (same or different alleles).

Although many of these points will have been covered during the teaching of mitosis and meiosis, students have a tendency to consider cell division purely in terms of chromosomes. Unless the relationship between chromosomes and genes is made explicit to them and the parallel behaviour of chromosomes and genes during cell division is noted, they are unlikely to recognise the relationship between cell division and inheritance. Because students also tend to see sexual reproduction in terms of physical activity rather than fusion of gametes to produce a new individual, they are often unaware that plants reproduce sexually. For this reason it is probably best, when first introducing inheritance, to avoid using plant examples. It is a good idea to begin by recapping the key points of cell division, noting what this means in terms of genes and genetic information. One approach would be to give an overview of the whole process that emphasises the continuity of information between generations. A possible structure for this is shown in Figure 7.2.

Figure 7.2 A summary of the process of inheritance

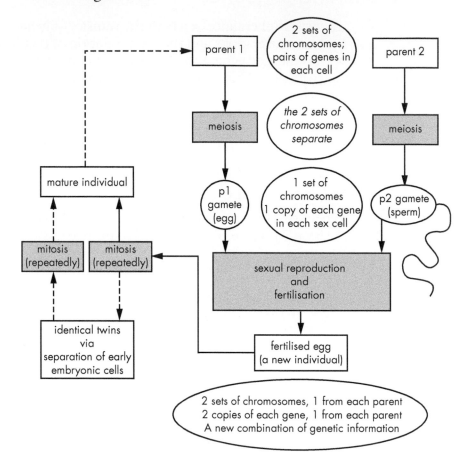

Key points when talking through a diagram like this might include the following:

- Meiosis has two functions – it halves the chromosome number and it increases variation.
- Each gamete is unique; it will have the same number of chromosomes (one set) and the same genes (one for each product) but no two gametes will have exactly the same combination of genetic information (alleles). For example, one gamete might contain the blue version of an eye colour gene and the brown version of a hair colour gene while another might contain the blue version of an eye colour gene but the blond version of a hair colour gene.

- Sexual reproduction is the mechanism by which genetic information from two different individuals can be combined to give a new and unique individual.
- Both parents make an equal contribution (matching sets of chromosomes).
- The genetic information in the newly fertilised egg is copied at each mitotic cell division so that each new cell has a complete copy of the information. In this way, every somatic cell in the new individual carries the same genetic information.

Animations of mitosis and meiosis are readily available on the internet and will help your students to understand the process.

Twins

Your students will almost certainly be interested in twins. They might like to know that non-identical twins are no more similar to each other, genetically, than to any of their other brothers or sisters who have the same father. This is because they have developed from two different eggs, each fertilised by a different sperm. The unusual thing is that their mother produced two egg cells at one time, instead of the more usual one. In contrast, identical twins both develop from the same egg, fertilised by a single sperm. In this case, the new embryo splits into two during an early (mitotic) cell division and each half then develops into a new individual. Because identical twins originate from just one cell they are genetically identical. Note that this does not mean that they will grow up to be the same in every way. Their characteristics will also be influenced by environmental factors (see Sections 7.4 and 7.6).

Having considered the general principles, you can use some specific examples to illustrate the ways in which genetic information will be interpreted within the new individual. This interpretation is at two levels – that of chromosomes and that of genes. Making clear the distinction between these two levels may help your students to develop their understanding of the relationship between genes and chromosomes.

■ Chromosomes and sex determination

It is easiest to start at the chromosome level, where interpretation is simpler and the objects can actually be seen during cell division. You might begin by using photos of human chromosomes (karyotypes) to illustrate the concept of matching sets of chromosomes. When looking at these it is helpful to remind students that chromosomes are normally enclosed within the nucleus and cannot be seen. They only become visible under the microscope during cell division, when the chromosomes become shorter and fatter, as they copy themselves, and are released into the cytoplasm as the nucleus breaks down. You can then go on to identify the sex chromosomes as being the odd pair out – they match (are homologous) in the female (XX) but do not match in the male (XY).

The link between inheritance and chromosome behaviour during meiosis can be made explicit by considering sex determination in humans. This also introduces the concepts of chance and probability. Key points that may need to be emphasised include the idea that:

- Pairs of chromosomes separate during meiosis, so that each sex cell has just one set of chromosomes; this includes the sex chromosomes.
- This means that in females X and X separate, so all eggs will carry an X chromosome, but in males X and Y separate, so half the sperm will carry an X chromosome and half will carry a Y chromosome.
- Because of this, it is the sperm that determines the sex of the child.
- Each fertilisation is an independent event; X and Y sperm have an equal chance of fertilising an egg; consequently the probability that any fertilisation will produce a boy (or a girl) will be 0.5.

You can help your students to understand these ideas using role play (see procedure below).

PROCEDURE

Sex determination role play
You will need: 1 neutral coloured card labelled: Egg (X); equal numbers of pink cards labelled: Sperm (X) and blue cards labelled: Sperm (Y); some willing students

1 Give one student the egg card and place them in one corner of the classroom: they are the egg, which has just been released from the ovary and entered the oviduct.
2 Hand out equal numbers of X and Y sperm, one per student: these are the sperm waiting in the testes, ready to go; group them at the other side of the classroom. Which one will fertilise the egg?
3 Random questions are used to simulate the effect of chance (for example, 'Are you wearing coloured socks?', 'Did you come to school on the bus?', 'Are you an Arsenal fan?', etc.); sperm can only move forward, one step at a time, when the answer is yes. The sperm will soon begin to spread out.
4 The sperm which reaches the egg first 'fertilises' it (they link arms and hold their cards up together) – is it a boy or a girl?

Students find the combination of probability, chance and independent events counter-intuitive. They feel, for example, that if a couple already have several boys, the next child *should* be a girl. They need to recognise that predictions based on probability are only meaningful at the population level and that chance effects can give a distorted impression in a small sample. For each fertilisation, the probability that the baby will be a girl is always 0.5 (a 1 in 2 chance) but because each fertilisation is an independent event (it is not influenced by the outcome of previous fertilisations) it is quite possible that, by chance, a couple will only have boys. However, when a large number of fertilisations are considered, the random effects of chance are balanced out. If all the births for a whole country are considered, then in any one year the number of baby boys will roughly equal the number of baby girls – that is, 50% of the babies will be boys and 50% will be girls and the proportion of boys to girls will be equal (1 : 1). You can illustrate this within the classroom by asking your students to toss a coin and record heads or tails. Continue until time or patience is exhausted then collate the results. For each toss there are only two possible outcomes (heads or tails) so the probability of either is 1 in 2 and the proportion of each should be 1 : 1. Within the class data there may be long sequences of either heads or tails but the larger the number of throws the more equal the number of heads and tails and the closer the result comes to the predicted ratio of 1 : 1.

■ Genes, alleles and phenotype

Things are a little more complex at the gene level. One reason for this is that pairs of alleles are not always identical – they may carry conflicting information. As a starting point here, it might be useful to go back to the example of fur colour in rabbits. Now that your students know that the gene for fur colour will be present twice in a rabbit, once in the set of chromosomes received from its father and once in the set received from its mother, do they see a problem? What happens if the rabbit receives one allele for black fur and one allele for white fur? In this case, even though the rabbit has alleles for both colours, it will have black fur. This is because the allele for black fur is dominant to the allele for white fur. The allele for white fur is recessive to the allele for black fur.

At this stage it might be appropriate to introduce some of the common terms and conventions that are used when discussing inheritance. Conventionally, the dominant allele is given a capital letter to represent its version of the characteristic (in this case 'B' for black fur) and the recessive allele is denoted by the lower case version of this same letter (in this case 'b' for white fur). It is then possible to describe the information in the genes (the genotype) using these letters as a shorthand. When a pair of genes carries the same information (BB or bb in this case) the organism is said to be *homozygous* for that characteristic. When a pair of genes carries different information (Bb in this case; conventionally, upper case letters come before lower case letters) an organism is said to be *heterozygous* for that characteristic. The physical expression of the genotype (black fur or white fur) is called the phenotype.

The relationship between pairs of alleles can be more complex than a simple dominant and recessive model would suggest. For example, there are three alleles in the ABO blood grouping system. Both A and B, when paired with O, are dominant, but when paired with each other they are co-dominant – neither dominates the other, they are both expressed in the phenotype and they are identified as a distinct blood group, AB.

Genes that are located on the X chromosome show atypical patterns of expression in males. This is because the X chromosome, and the genes located on it, usually occur only once in male cells. As a result, the alleles remain unpaired and each one will be expressed in the phenotype. The important point here is that a single allele cannot be dominant, recessive or even co-dominant. These terms are used to describe the relationship between *two* alleles, in a way that helps us to make sense of the resulting phenotype and make predictions about the inheritance of that characteristic. In theory we might also expect to see some Y-linked inheritance, since genes on the Y chromosome only occur once and their allele remains unpaired, but very few have been identified. Many students have difficulty with this concept of sex linkage. This may be because they don't realise that the terms 'dominant' and 'recessive' apply to a particular relationship, rather than a particular allele. It may also reflect their general uncertainty about the relationship between genes and chromosomes. The following activity might help your students to understand these interrelationships better.

PROCEDURE

Modelling chromosomes, genes and alleles
You will need (as a minimum): a small hat, and a larger hat that can cover the smaller one; a small/thin scarf and a large scarf that can cover the smaller one; a pair of gloves or mittens (preferably the sort that will fit either hand); a bag to put these accessories in. You will also need two students of the same sex and roughly the same size.
Note: the more outlandish the above props, the more memorable the activity.
The basic activity:

1 Bring the two students to the front of the class, stand them side by side and introduce them – they are a pair of homologous chromosomes.
2 Invite them each to pull a hat out of the bag and put it on – introduce the 'hat' gene; note that it is in the same place on each chromosome, that it is easily recognisable as a hat, but that it occurs in two different forms (alleles).
3 Repeat this for the scarves.
4 Ask the class which 'hat' allele is dominant (would appear in the phenotype) – confirm this by putting the dominant hat on top of the recessive hat.
5 Repeat for the scarf gene.
6 Invite the students to pull a glove out of the bag and put it on – introduce the glove gene and note that in this case both chromosomes have the same allele, so there is no choice about the phenotype.

(contd)

Extension of this activity:

1 Use the three 'genes' from the basic activity to re-enforce the terminology: identify/record the genotypes, note whether they are homozygous or heterozygous.
2 Extend the number of alleles and the possible relationships:
   a Add a middling sized hat to the mix (this will be dominant to the small hat but recessive to the big hat).
   b Add a second 'glove' allele – some large fingerless gloves; is either of the two alleles dominant? The original glove allele must go under the fingerless gloves but their fingers will continue to poke through so both alleles will be visible in the phenotype; they are co-dominant.
   c Add more 'genes' – for the fun of it.

[adapted from Dolan, 1996]

■ Single gene (monohybrid) inheritance

With some awareness of how genetic information is passed on to the new individual, and some awareness of the relationship between alleles, your students should now be ready to consider monohybrid inheritance in more detail and to start using the Punnett square. You could begin by using the rabbit fur example and then move on to consider other examples, including single gene disorders in humans.

In rabbits

When a homozygous black rabbit (BB) is crossed with a homozygous white rabbit (bb), all the offspring are heterozygous black rabbits (Bb) – the white colour appears to have been lost. If this first generation is allowed to interbreed, the white colour will reappear again in their offspring. In this second generation, the genotypes will be present in the ratio of 1 homozygous black (BB) to 2 heterozygous black (Bb) to 1 homozygous white (bb), giving a phenotypic ratio of 3 black to 1 white rabbit. The probability that any one rabbit in the second generation will be black is therefore 0.75. This can be shown using a Punnett square. Your class is likely to divide into those who find it easy to use a Punnett square and those who struggle to understand what is going on. This second group will need support, but you shouldn't feel complacent about the first group. There is ample evidence that many students can learn the technique of completing Punnett squares without understanding the processes or outcomes that these squares represent. Such students will happily produce a Punnett square to show the outcome of sexual reproduction between smooth and wrinkled peas, even when they don't believe that plants are capable of sexual reproduction. You can help your students to visualise the process by talking through the different stages and representing each one diagrammatically, using chromosomes as well as alleles, before completing the Punnett square. Figure 7.3 can be used as a template for this.

Figure 7.3
Drawing up a
Punnett square

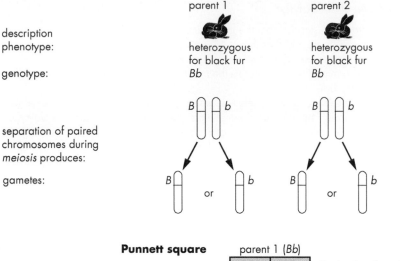

description
phenotype:

parent 1

parent 2

heterozygous
for black fur
Bb

heterozygous
for black fur
Bb

genotype:

separation of paired
chromosomes during
meiosis produces:

gametes:

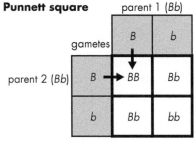

Punnett square

Each white box shows
one of the four possible
outcomes at fertilisation:

1 × BB   2 × Bb   1 × bb

(3 black rabbits and 1
white rabbit).

In humans

A relatively small number of human characteristics and genetic
disorders are caused (or behave as if they are caused) by variations
in a single gene but these are the ones most commonly referred to
in textbooks since these are the ones that are easiest to interpret and
predict. They will catch the interest of your students but they need
to be handled sensitively.

If you are using eye colour as an example, treating it as a single
gene characteristic with B representing the brown allele and
b representing the blue allele, some students may apply the lesson to
their own family and learn more than you intended. If both parents
have blue eyes but your student has brown eyes s/he might start
asking some difficult questions. The possibility of this happening
may be reduced by emphasising that a description of the inheritance
of eye colour based on a single gene and only two alleles, blue and
brown, is a simplified explanation. While it is accurate as far as it goes
– brown is dominant to blue – other genes and other colours, such as
green, are also involved and the interrelationship between the
different factors is complex. It is also worth remembering that
variation within a gene arises from mutation so this particular brown
allele might have arisen by a chance mutation (see Section 7.4).

A number of genetic diseases follow a monohybrid pattern of inheritance (see Table 7.2 for some examples) and will be of considerable interest to your students but many of them have serious, sometimes fatal, consequences. Before going into details about any of these diseases, it is a good idea to probe the prior knowledge and experiences of your students. Have they heard of the disease? If so, what do they know about it and how did they come to hear of it? You may also want to consider the whole school situation – does any student in your school have that particular illness? You can then adjust your approach, or modify the information you choose to give, accordingly. For example, the continuous respiratory and digestive problems that are symptomatic of cystic fibrosis put a strain on the whole body and reduce the life expectancy of those affected. While you might want to discuss the basic symptoms of cystic fibrosis, and their treatment, you might prefer to leave out the consequences (reduced life expectancy) – especially if any students, or their family or friends, are affected.

Table 7.2 Examples of inherited single gene human disorders

Disorder	Description	Inheritance
cystic fibrosis	excess mucus causing respiratory problems and poor digestion	recessive
thalassemia	accelerated destruction of red blood cells	recessive
sickle cell anaemia	malformed red blood cells lead to reduced uptake of oxygen and excessive wear and tear	co-dominant: heterozygous individuals have a mix of normal and sickle red blood cells
Huntington's disease	progressive degeneration of the nervous system; no symptoms until late 30s or 40s	dominant
haemophilia	lack of blood-clotting factors leading to excessive bleeding from wounds	recessive but sex linked

For many genetic diseases there seems to be a relationship between phenotype and environmental conditions. As a result they are often associated with particular geographical areas and show racial differences. The most common genetic disease amongst Northern Europeans is cystic fibrosis, thalassemia occurs more commonly in people originating from Mediterranean countries and certain regions of India and sickle cell anaemia occurs most commonly in people originating from Africa and the West Indies. The persistence of these potentially lethal diseases in a relatively high proportion of these populations suggests that they confer (or conferred in the recent past) some selective advantage in the heterozygous form. This has been shown for sickle cell anaemia, where heterozygotes have an increased resistance to malaria (see Section 7.4). However, singling out just one disorder for further discussion may give the unintended impression that the affected groups are somehow genetically inferior to other groups. If you only have time to consider one disease then it may be best to focus on the one that is most significant for your student group. In most parts of the UK this would probably be cystic fibrosis, since one in 27 people in the UK are carriers of this disease. That is, they are heterozygous for a cystic fibrosis allele and can pass the disease on to their children, even though they themselves are unaffected. Because heterozygotes are symptomless they may remain unaware of their carrier status unless their partner is also a carrier and they have an affected child (see Section 7.5, Genetic screening).

■ Discrete (single gene) versus continuous (multifactorial) characteristics

Each of the inherited characteristics mentioned so far is controlled by one, or a very few, genes. These represent the minority of genetic characteristics, but they are the easiest to explain. Most characteristics are multifactorial – they result from a combination of many genes, and environment factors, interacting. For example, height is in part inherited and in part determined by environmental factors such as diet.

You can demonstrate the difference between discrete (single gene) characteristics and continuous (multifactorial) characteristics by looking at variation within your class. Collect data on a range of characteristics from the whole class and ensure that it includes some continuous characteristics such as height (if you have already collected data on discrete characteristics you will only need to collect data on continuous characteristics now). It may be best to avoid weight as many teenagers are particularly sensitive about this. Separate the characteristics into those for which there are a discrete number of independent variations, e.g. eye colour, and those that show a continuous range of values, e.g. height. For a discrete characteristic, the frequency of different variations within the population can be shown using a bar chart. For a continuous characteristic, the frequency of different values can be shown using a histogram and the result, if enough data are collected and an appropriate interval selected, will be a bell curve or 'normal distribution'. You can increase the effectiveness of this activity by setting up a database and, each year, asking your students to add their data to it. As your population size grows, the normal distribution will become clearer. It is important to remember that some characteristics are caused entirely by the environment, e.g. scarring as a result of injury, or the docked tails of some domestic animals.

Enhancement ideas

■ Chromosomes

- Humans normally have 23 pairs of chromosomes (46 chromosomes in total) but occasionally a person has one extra or one less chromosome. For example:
  - Turner's syndrome (XO – a female with only one X chromosome)
  - Kleinfelter's syndrome (XXY – a male with an extra X chromosome)
  - Down's syndrome (usually an extra chromosome 21).

  Only a few such conditions are known to occur. This is because errors at the chromosome level always involve a large number of genes and usually result in major developmental problems that are incompatible with life. In most cases the fertilised egg fails to develop beyond the early stages of pregnancy. It is interesting to note that too much genetic information seems to be as damaging as too little genetic information.
  Students could find out more about these (and other) chromosomal conditions, think about how they might arise (non-disjunction of chromosome pairs during meiosis) and work out patterns of inheritance.

- If too much genetic information can be damaging, and males only need one X chromosome, your students may recognise that having two X chromosomes could be a problem. Exploring this issue could make a small but interesting student project, perhaps starting with the question 'why, if additional genetic information is potentially life threatening, do people with Kleinfelter's syndrome have few if any symptoms?' They could then be encouraged to go beyond the idea of random inactivation of one X chromosome in each cell and consider the process and possible consequences such as 'mosaic' phenotypes. An example of this is tortoiseshell fur colouring in cats (see also Section 7.6) – one of the genes for fur colour is found on the X chromosome and comes in two forms, orange or black/dark brown; if a cat is heterozygous for this gene each cell will have a black/dark brown allele on one X chromosome and an orange allele on the other; because X inactivation is random and happens at an early stage of development cells with an active orange allele give rise to orange patches of fur and cells with an active black/dark brown allele give rise to a patch of black/dark brown fur; the result is a random mosaic of orange and black/dark brown fur which is the typical tortoiseshell colouring.

■ Inheritance

- Heterozygous and homozygous dominant individuals are usually indistinguishable, phenotypically. It is possible to identify heterozygotes by a process of back crossing (or 'test' crossing). Using the rabbits as an example, it is possible to identify the heterozygotes by mating the black rabbits (BB or Bb) with white rabbits (homozygous recessive, bb). If the black rabbit is homozygous (BB) then all the offspring will be black. If the black rabbit is heterozygous (Bb) then, on average, half the offspring will be black and half will be white. Your students might enjoy problem solving which uses this knowledge.
- Back crossing is obviously not possible when looking at human inheritance. Before genetic screening became available (see Section 7.5) heterozygotes for a particular condition or characteristic were identified by pedigree analysis and a process of deduction. For example, if two apparently unaffected people give birth to a child with cystic fibrosis then, assuming the cause isn't a mutation, they must both be heterozygous for cystic fibrosis; it is then possible to work backwards from these parents to identify other carriers. Some of your students will enjoy trying to work out the genotypes for different members of a family, based on their phenotypes.
- Students might also like to explore historical examples of inherited conditions such as King George III and porphyria and Queen Victoria and haemophilia.

7.3 Gene expression

Previous knowledge and experience

In preparation for this section it will be helpful if your students:

- have a basic understanding of genes and the relationship between DNA, genes, chromosomes and cells (as in Section 7.1)
- understand the purpose of mitosis: to provide an organism with new cells for growth and maintenance (replacement of dead/ damaged cells)

- understand the basic process of mitosis and the consequence: that an exact copy of all the DNA is made at each mitotic cell division so that each new cell has a complete copy; as a result, all cells within an organism carry the same genetic information (exception: gametes, which are produced by a different kind of cell division – meiosis).

A teaching sequence

If your students are to have any understanding of gene technology, to make any sense of media reports on new developments in genetics or to have informed discussions of related issues, they need some understanding of the processes by which information encoded in the DNA is translated into the diversity of materials that enable an organism to function. These processes are complex but the basic principles can be made accessible to most students if non-essential details are kept to a minimum.

■ DNA: the basic structure

If they are to develop some understanding of gene expression your students need to know something about the structure of DNA. As a minimum your students need to know that:

- DNA is made up of two long strands, held together by pairs of bases.
- There are only four kinds of base (A, T, C, G).
- These bases pair in a very specific way (A only pairs with T; C only pairs with G).

Getting your students to build simple models of DNA, during which they have to match the bases, is probably the most effective way of explaining this very basic structure. See the activity below for a particularly popular version. Or visit www.yourgenome.org/downloads/yummy_gummy_inst_A4.pdf for a similar method. It can also be exciting to be able to see DNA – the second activity gives a simple method for extracting DNA based on household ingredients.

PROCEDURE

Using sweets to model DNA

You will need: long candy or liquorice cords with fondant centres, all the same colour (two per model); large bags of Liquorice Allsorts or Dolly Mixtures (at least 300 grams in total); cocktail sticks (at least 20 per model); a good reason why your students should not eat the sweets!

Building the model:

1 Identify four different types/colours of sweet to represent the four base pairs (A, T, C and G).
2 Pick out ten of these 'bases' and attach them to a liquorice 'backbone' at regular intervals using the cocktail sticks.
3 Find the corresponding base pairs.
4 Connect each base on the backbone to its corresponding base using a second cocktail stick.
5 Push this second cocktail stick right through the corresponding base and into the second liquorice 'backbone'.
6 You should now have a double-stranded piece of 'DNA' with ten base pairs.

A range of similar models, using a variety of materials, can be found on the internet.

PROCEDURE

DNA extraction using household materials

You will need: a 500 cm³ bottle of methylated spirits in a bowl of ice (~100 cm³ per extraction); 25 g salt and 80 g cheap (less concentrated) washing up liquid dissolved in 900 cm³ water; some kiwi fruits (1 per extraction), peeled and mashed; a pan or water bath of hot (50–60 °C) water; a sieve, a small bowl, a measuring cylinder (100 cm³) and a glass or beaker (about 200 cm³).

Instructions

Some people are allergic to kiwi fruit. Take care when handling large amounts of methylated spirits.

1 Put the mashed kiwi fruit into the small bowl and add 100 cm³ of the salt mix.
2 Give this mix a stir then place it in the hot water for 15 minutes.
3 Sieve this warm mix into the glass or beaker.
4 Gently pour ice cold methylated spirits down the side of the glass until there are equal volumes of green kiwi and purple meths.
5 Where the two layers meet, a white stringy layer should form – kiwi DNA!

For best results:

● Mash the kiwi fruit as much as possible.
● Leave the mix in the hot water for as long as possible.
● Keep the methylated spirits as cold as possible until the last possible moment (but don't put it in the fridge unless you have one that's designed for use in a laboratory).

For more details and additional information, see www.york.ac.uk/res/sots/activities/diydna.htm.

■ DNA: coding the information

One way of explaining how DNA can provide all the genetic information that an organism needs is to use a language analogy in which DNA is a book, genetic information is the text, genes are the sentences and the genetic code is the language. Using this analogy you can explain the following:

- This particular language has only four letters (the four DNA bases).
- Each 'word' is very short – just three letters (a triplet of bases).
- There are 64 ways in which four letters can be rearranged into groups of three, so this 'language' (the genetic code) is very simple, it uses a 'vocabulary' of just 64 words.
- A near infinite number of different messages can be written with a vocabulary of 64 words.
- The sequence of words within each 'sentence' (gene) provides instructions for making the products associated with that gene.
- Every living thing uses the same 'language' (the same genetic code) but different types of organism use it to produce different stories (sets of instructions) resulting in very different characteristics.
- In preparation for Section 7.5 you might also want to note that because all living things use the same 'language' (genetic code), every organism can read and understand the 'text' (genetic information) from any other organism – this is what makes genetic engineering possible.

These last two points are very important. Many students are confused between the genetic code (the 'language' of genetics) and the message that has been written with that code (the 'text'). Some students find it almost impossible to accept that the same code is used by all organisms.

■ Reading the message

Even if your students have understood the basic concept of inheritance (that there has been a physical transfer of information from parent to offspring) and also understood that as this fertilised egg divides and the new organism grows, this new combination of genetic information is accurately copied to all new cells, they may still have a problem. If all somatic cells within an individual contain the same genetic information, then how come different types of cell look so different? Intuitively, they feel that each type of cell must contain just those genes needed for its particular function. You can check the ideas within your class by asking your students to compare the genetic information found in different types of cell taken from the same individual (see Worksheet 7.2). As with Worksheet 7.1, you will learn most about your students' thinking if you probe their reasoning.

Worksheet 7.2 Probing students' understanding of gene expression

Cells	Same or different genetic information?	Reasons
Two cheek cells from the same individual		
One cheek cell and one nerve cell from the same individual		
One cheek cell and one sperm cell from the same individual		
Two sperm cells from the same individual		
Two cheek cells from two different individuals		

The final comparison on this worksheet is between cheek cells from two different individuals. At this point students often become aware of a contradiction in their thinking – that two cheek cells should have the same genetic information because they do the same job but cells from two different individuals will carry different information because all individuals (except identical twins) are genetically unique. Exploration of this contradictory thinking can help students to develop a more coherent understanding of the science.

What many students don't realise is that genes can be switched on and off in response to their environment. This means that in any one cell, at any given time, only a small proportion of genes will be active. This needs to be made explicit and a familiar example might help: sunshine contains potentially harmful UV light; skin cells respond to sunlight by producing a pigment, melanin, which protects the underlying cells from the UV light; this is very apparent in people with pale skin – they become suntanned; in the winter months there is little sunshine; in the absence of sunshine the genes that produce melanin are switched off and the suntan fades.

Enhancement ideas

■ DNA structure

- You could provide additional information about the structure of DNA: the backbone is made up of alternating sugar and phosphate groups; the bases are attached to the sugar in the sugar/phosphate 'backbone'; one base plus one sugar group plus one phosphate group is a nucleotide; adenine and guanine (purines) are bigger than thymine and cytosine (pyrimidines); a purine always pairs with a pyrimidine; the specificity of pairing is linked to the weak hydrogen bonding between the bases: the A–T pair has two bonds and the G–C pair has three, etc.

- Your students could develop more sophisticated models of DNA to represent this additional detail – by refining the sweet model, using molecular models such as molymod (www.molymod.com/) or paper models (see Origami DNA at www.yourgenome.org/downloads/activities.shtml) or making their own. Keep these models – you might want your students to use them to extend their understanding of gene expression.

■ Gene expression

- Your students could use the DNA models (perhaps linking several together) to look in more detail at how these 64 'words' provide information:

  - Pull some of the base pairs apart to 'unzip' one end of the DNA model (at least nine base pairs).
  - Build a complementary strand of 'messenger' RNA (mRNA), remembering that thymine is replaced with uracil in RNA.
  - 'Detach' this single strand of mRNA from the DNA and move it away (in the cell it would move out of the nucleus into the cytoplasm); re-zip the DNA strands.
  - Identify each triplet on the RNA strand, in sequence, and use the genetic code (found in most A level textbooks) to identify the corresponding sequence of amino acids; alternatively use the codon wheel available at: www.yourgenome.org/downloads/function_finder_forweb.pdf
  - Write out the sequence of amino acids that the DNA model/RNA strand specifies.

In doing this students will probably discover that some words are used as punctuation (start and stop signals) and that there is more than one word/code for each amino acid – can they see any advantages in this?

- Additional information they might find useful or interesting could include: amino acids join up to form polypeptide chains and proteins are formed from one or more polypeptide chains; the mRNA sequence may be 'edited' before it is used to make a polypeptide chain (for example, some short sequences may be cut out); there are other types of RNA (transfer RNA, ribosomal RNA) and sometimes these, rather than proteins, are the gene product.
- Any change in the sequence of DNA bases (the sequence of letters) has the potential to change the squence of amino acids in the polypeptide chain (the meaning of the sentence). For example, in sickle cell anaemia a small change in the DNA sequence changes just one amino acid in the haemoglobin molecule – a glutamic acid is replaced by a valine. This one change alters the way in which the haemoglobin responds to oxygen and so causes the disease. In preparation for Section 7.4 you could use the DNA model to look at mutations. Unzip the double-stranded DNA as before but this time when building up the complementary strand of mRNA make a mistake – an incorrect pairing or a missed pairing. Check – does this change the final amino acid sequence? If not, your students might want to reconsider the advantages of having multiple codes for a single amino acid.

7.4 The genetic basis of variation

Previous knowledge and experience

The concept of variation, both in terms of the diversity of life and the uniqueness of individuals, should be well established. To appreciate the causes and consequences of variation your students will need:

- knowledge of the physical process of meiosis (see Chapter 1)
- an awareness that DNA is the source of genetic information
- a basic understanding of inheritance.

This section focuses on the genetic basis of variation – the role of sexual reproduction and the implications for gene frequencies within the population as a whole (the gene pool). The ideas are relatively complex and may be better suited to more advanced students. For a more general view, with a greater emphasis on environmental factors, see Chapter 8. Mutations within somatic (body) cells are considered in Section 7.6.

A teaching sequence

Within one species there is a considerable degree of variation between individuals. Even within one family there is variation. Where does this variation come from?

The process of sexual reproduction provides several opportunities to mix genes and their alleles in new combinations and so increase variation between individuals. As we have seen (Section 7.2) the somatic cells in an individual carry two sets of chromosomes, one set from each parent. Each pair of chromosomes carries the same set of genes but each pair of genes may carry slightly different information (the alleles may be the same or different).

During the process of meiosis (see Chapter 1), in which the number of chromosomes is halved, random assortment of chromosomes mixes maternal and paternal alleles from different chromosomes and crossing over mixes maternal and paternal genes within a chromosome. The net result of these two processes is that no two eggs or sperm will have the same mix of alleles. The random nature of fertilisation increases variation even further and ensures that no two offspring of the same parents will have the same combination of alleles (with the exception of identical twins).

This shuffling of alleles can rearrange existing alleles into new combinations but it cannot change the total amount of genetic variation within the population. A new variation can only arise within a population if the DNA 'text' within the gametes is changed by a mutation. This is a relatively rare event. You can illustrate the effect of a mutation using a simple sentence, e.g. 'The mat was wet'. If one letter is substituted (analogous to changing one base pair in the DNA) this becomes 'The man was wet'. The words still make sense but the meaning is quite different (see Section 7.3, enhancement ideas). A number of environmental factors will increase the rate of mutation, including ultraviolet light, X-rays and carcinogens such as tobacco. Although mutations can occur in any cell (somatic cells or gametes) mutations occurring in somatic cells do not increase the amount of variation in the species as they are not passed on to subsequent generations. Only mutations occurring in the gametes have the chance of being passed on to offspring and so entering the gene pool.

A common misunderstanding about mutations is that they are always a bad thing. In practice, mutations are neither bad nor good. They can only be defined in this way by considering the effect they have on the organism in which they occur and this will usually depend on environmental factors. At some time in the past a mutation occurred in the peppered moth (*Biston betularia*), which resulted in the occurrence of two distinct forms – a pale form and a melanic (dark) form. When the countryside was relatively unpolluted, the pale form was the more common. The dark form was very conspicuous against the pale bark of the trees and was quickly eaten by predators. As a result, very few dark moths survived long enough to reproduce and pass the dark allele on to their offspring. With industrialisation the countryside became polluted and tree bark darkened. The pale moths were now conspicuous and quickly eaten by predators while the dark moths survived and reproduced. The mutation that had once been a disadvantage had become an advantage and the dark form became the more common form of this moth. Over the years the original data and conclusions have been contested and alternative explanations have been presented (this might provide a useful case study for How Science Works) but the peppered moth remains a helpful illustration of how a change in the environment can change the impact of a particular mutation.

In general, any mutation that gives an organism a selective advantage over the rest of the population will increase in frequency within that population, and any mutation that results in a selective disadvantage will decrease in frequency within that population. The consequences of this, in terms of evolution, are discussed in Chapter 8. Occasionally, a mutation that appears to confer a disadvantage is maintained at unexpectedly high levels within a population. In some cases this is because the heterozygote confers some selective advantage, as is the case with sickle cell anaemia (see Section 7.3). In other cases this is because the characteristic that causes the disadvantage only becomes apparent in the mature individual. By this stage reproduction may be complete and the affected gene will already have been passed on to the next generation. An example of this is Huntington's disease.

Many mutations confer neither advantage nor disadvantage and are simply maintained within the population as one of the variations.

7.5 Genetic technologies

Previous knowledge and experience

Understanding the basic principles of these techniques will require a good understanding of the basic genetic concepts introduced in Sections 7.1 and 7.3.

If your students are to engage in reasoned discussion of the social and ethical consequences of gene technology they will also need some understanding of the basic concepts presented in Section 7.2. What they will not need is a detailed understanding of the techniques or the genetic concepts. This means that all students, regardless of academic ability, can engage in such discussions.

A teaching sequence

When techniques for manipulating DNA were first developed, the focus was on manipulating the whole genome (e.g. DNA fingerprinting), identifying useful or important genes and their alleles (e.g. genetic screening) and manipulating genes (e.g. genetic engineering), with a strong emphasis on single gene disorders and characteristics. Alongside this went an interest in gene sequencing – identifying the sequence of bases within a gene. Initially the process of gene sequencing was very labour intensive but with automation the sequencing of whole genomes became a reality.

The scale of the Human Genome Project (HGP) provided an unusually public insight to the fascinating and contradictory mix which was science in the making and it might make a good case study for How Science Works – the co-operation between research groups; the power clashes and sharp practices between vested interests (whether for personal reputation or commercial gain); the tension between public benefits and commercial restrictions, etc. One reason for the intensity of competing interests was the expectation that many useful genes, each linked to a specific characteristic, would be discovered. For some, the reality turned out to be a disappointment. On completion, the HGP showed that most characteristics are determined by a complex web of interactions between multiple genes and their environments and very few new examples of characteristics determined by a single gene were found. The result has been a quiet revolution in our thinking about genetics, which has far-reaching implications – for science and for society. For many researchers the purpose and focus of their work has changed from hunting for potentially useful genes to working with whole genomes, trawling through DNA databases looking for potentially interesting patterns or relationships. While this section looks at 'traditional' gene technologies, Section 7.6 focuses on current areas of interest and genomics in general.

The astonishing pace of developments in genetics and genomics can sometimes make it difficult for your students to distinguish between fact and fiction, especially when fiction (science fiction, soap operas and police dramas) rather than factual news or science books is their main source of information.

A class discussion focusing on video clips from a popular TV programme, or on newspaper cuttings about some recent development or application of genomics, might be a useful way into this topic. You could use this discussion to sound out your students' knowledge and understanding of different gene technologies and their ability to evaluate the potential of gene technology – what is, and what is not, scientifically feasible. It might also reveal some of the misunderstanding that your students have about the underlying genetics. Some suggestions for how to do this can be found under 'Social and ethical implications' on page 206.

Some curricula and syllabuses make specific demands regarding the techniques that must be covered and simple details of the main techniques are given below. In preparing to teach some of the basic principles, you might find it helpful to identify the genetic concepts on which they are based and to consider the extent to which your students understand these concepts. You can then help your students either by simplifying the basic principles or by providing additional information about the genetic concepts. It's probably best to treat discussion of DNA sequencing as an enhancement activity.

■ DNA fingerprinting

In the technique of DNA fingerprinting, samples of DNA are cut into fragments using restriction enzymes. The fragments are then separated by size, using gel electrophoresis, and stained to produce a pattern of bands – the fingerprint. Because no two individuals have identical DNA (possible exception: identical twins) and each restriction enzyme is highly specific in its action, only cutting the DNA when it encounters a very particular sequence of base pairs, every person will produce a different genetic fingerprint and the probability of any two individuals producing the same pattern of bands is close to zero. This technique can be used in:

- forensic work – to check DNA from suspects against DNA from the scene of the crime, looking for a match
- paternity cases – to compare the DNA of the child with the DNA of the putative father; the child receives half of its genetic information from its father so half of the bands in the child should match half of the bands in the father
- preservation of wildlife – for example, to confirm the origins of captive birds and identify those taken illegally from the wild.

Students are sometimes confused between DNA fingerprinting and traditional fingerprinting.

■ Genetic screening

Genetic screening is used to test a sample of DNA for the presence of a specific allele. Probes carrying a sequence of bases that are complementary to the allele of interest are mixed with the DNA sample. If the allele is present in the DNA the probes will stick to it, acting as an identifiable marker. Screening is available for a number of genetic disorders including cystic fibrosis and Huntington's disease. The information it provides can help people to make informed decisions about their life, including whether or not to proceed with a pregnancy, but there is always a small risk of a false positive or a false negative result. Some tests are available commercially, through mail order.

Screening raises a number of social, ethical and moral issues, including confidentiality – who should, and who should not, have access to the information?

■ Cloning

Cloning can apply to the whole genome (producing new cells or organisms that are genetically identical to the original) or to an individual gene.

Cloning the whole genome can occur naturally, for example: the production of identical twins when cells from one fertilised egg are separated early in development; the regeneration of an organism from bits that have split from an existing organism (primitive invertebrates, many plants); asexual reproduction (bacteria, plants). Some cloning techniques are based on these naturally occurring processes, for example: separating the cells in unspecialised tissue and growing them on; making leaf cuttings from plants, etc. Some techniques are more complex, for example nuclear transplantation (taking the nucleus from an egg and replacing it with one from a somatic cell – the technique used to produce Dolly the sheep). Within an organism cells are cloned naturally through mitotic cell division but cell lines can also be maintained by growing cells in culture in the laboratory, in much the same way that bacteria are grown. This is usually done when the genome of the cells is useful in some particular way. For example the HeLa cell line contains a cell, originally removed from a patient with cancer, which grows and divides very rapidly. This was the first cell line to be developed and was originally made freely available to any researcher who wished to use it, but the patient and their relatives were never asked for permission and this raised major ethical issues once commercial interests became involved.

Cloning genes usually involves the use of the polymerase chain reaction (PCR). The sequence of DNA containing the required gene is identified and cut out, added to a chemical soup of nucleotides and polymerase enzymes and taken through several cycles of heating and cooling. As the mixture heats up the double strands of DNA separate, resulting in single-stranded DNA; as the mixture cools down a complementary strand of DNA forms against each single strand; as a result, with each cycle each double-stranded DNA forms two new double strands and there is an exponential increase in the number of copies. Animations such as www.maxanim.com/genetics/ PCR/PCR.htm can be used to illustrate this.

■ Genetic engineering

Genetic engineering means adding, deleting or transferring genetic material to a genome and can involve organisms of the same species or organisms of quite different species. A find–cut–copy–paste technique is used, analogous to word processing. The required gene is identified, cut out of the DNA using restriction enzymes and cloned (see 'cloning', above) to produce thousands of copies. Copies are then inserted into the cells of the host organism, together with a switching mechanism to ensure that the gene can be turned on. The uses to which genetic engineering can be put are almost limitless and include:

- the production of transgenic organisms (bacteria, sheep) that can produce human proteins
- gene therapy – to replace a non-functional allele with the functional version, e.g. to relieve the symptoms of severe combined immunodeficiency (SCID) or cystic fibrosis
- agriculture, e.g. to increase the yield or improve pest resistance in crops (resulting in genetically modified food).

It is sometimes said that genetic engineering is no different from selective breeding, a process that humans have been engaged in for thousands of years. This is not necessarily true. In selective breeding we select two individuals of the same species who have some desirable characteristic and we breed them in the hope of getting more individuals with the same desirable characteristics. By applying selective pressure in this way over many generations we can produce breeds or varieties with very distinctive characteristics, e.g. lean cattle, high yield crops, fast race horses. Genetic engineering between two organisms of the same species *is* directly comparable to this process. The advantage is that genetic engineering speeds up the process and removes the element of chance associated with sexual reproduction and inheritance. However, genetic engineering between two individuals from different species is *not* comparable to selective breeding. No amount of selective breeding would ever produce a cow whose genome included the human gene for insulin, or a plant containing an 'anti-freeze' gene from a fish. Students are sometimes confused between selective breeding and genetic engineering.

■ DNA sequencing

DNA sequencing uses a similar technique to gene cloning but in addition to the usual mix of nucleotides and polymerase enzymes the chemical soup also contains 'terminator' nucleotides (A, T, C and G). As the ordinary nucleotides pair up against the single-stranded DNA, the sugar of one reacts with the phosphate of the next and the sugar/phosphate backbone of the complementary strand of DNA is formed. The terminator nucleotides are modified so that when they pair up against the single-stranded DNA, their sugar is unable to react with the phosphate on the next nucleotide, so synthesis of the complementary strand of DNA is terminated. Through many PCR cycles complementary strands of DNA of different lengths will be produced, depending on where the terminator nucleotide became attached. These are separated by size using gel electrophoresis (for small lengths of DNA) or capillary electrophoresis (for automated sequencing of extended lengths of DNA) and because each type of terminator nucleotide carries a different coloured fluorescent marker which identifies their base (A, T, C or G) the sequence of bases on the DNA can be identified (see Figure 7.4).

Figure 7.4 DNA sequencing – interpreting the gel. Note: the smallest molecule of DNA is at the top of the gel as it moves fastest

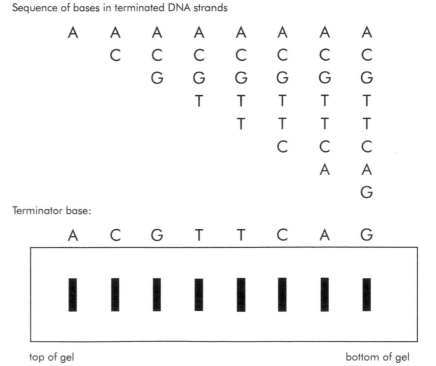

See www.yourgenome.org/teachers/sequencing.shtml for an animation of automated DNA sequencing.

■ Social and ethical implications

Gene technology makes use of a number of different techniques and appears to have almost limitless potential. Because it involves the manipulation of DNA, which is the very basis of life, the use of gene technology raises a number of important social and ethical issues. It is likely that all your students, at some point in their lives, will be faced with decisions relating to the use of gene technology – for example, whether or not to undergo genetic screening for particular inherited disorders. Consequently, a discussion of gene technology and its uses can have real relevance for them and help to stimulate their interest in the underlying science. It will also give them valuable experience of discussing social issues and evaluating information and ideas within a scientific context. Before starting such discussions it is a good idea to be very clear in your own mind about what it is that you want your students to have learnt by the end of it – to present a reasoned argument, to justify a claim or point of view, to evaluate evidence, to assess risk, etc. This will enable you to manage the activity to maximum effect and provide additional support or guidance if needed, for example: prompting students explicitly to justify their point of view; providing criteria they can use to evaluate the quality of the evidence (Is the information in a popular newspaper or a scientific journal? How reliable are the data? Was the design of the experiment valid? etc.), and so on. Guidance on the management of such discussions can be found on some websites, for example www.at-bristol.org.uk/cz/teachers/czhowto_web.pdf and www.windfalldigital.com/ethicalemporium/.

To encourage students to come to a view that they can justify, through exploring and evaluating their own ideas and considering the views of others, it is best to focus the discussions on a specific use of gene technology. Your students will not need a detailed understanding of the science in order to engage with the issues and come to a view, but they will need a brief outline of the context. You can focus the discussion further by providing a number of questions for your students to answer. An example is given in Figure 7.5. If your students are unfamiliar with the context they may find it difficult, at this stage, to identify the important issues. This will limit the discussion. You can help by mentioning some of the key issues.

Working in groups of four will give all your students an opportunity to express their views, but they will need some ground rules. In particular, everyone should have the opportunity to contribute to the discussion and everyone should treat the views of the others with respect, even when they disagree.

Figure 7.5
Providing a focus
for a discussion of
gene technology

!!!STOP PRESS!!!

**And You Thought
It Was All In Your Jeans!**

A London clinic has recently been offering to produce 'Designer Babies' for parents. For just £50 000 the clinic will check and, if necessary, change the parents' genes in order to produce the baby of their choice. Once selected, the baby develops normally inside the mother. The choice at the moment is limited to sex, intelligence, height and hair colour, but a spokesman said that several other features would soon be available. All 'Designer Babies' are guaranteed free from identifiable genetic diseases.

Asking each group to try to reach a consensus is a good way of encouraging your students to articulate their thinking and to justify their points of view, but don't necessarily expect them to achieve it! As there may be no right or wrong answers in this activity, just a diversity of views which have been justified to a greater or lesser extent, it is unlikely that they will manage to agree completely. Feedback from each group to the whole class at the end of the discussions will allow the full range of views within the class to be shared by everyone, and perhaps widen the debate. Few students will have difficulty giving an opinion, so this activity is suited to all ability levels. What they find more difficult, at all levels, is to justify their ideas (to explain why they think what they think) and to evaluate different ideas. Some students will need support in doing this.

In some cases a lack of understanding of the science may mean that your students do not recognise an important issue. For example, students who do not differentiate between gametes and somatic cells will see little difference between somatic gene therapy and germ-line gene therapy and are likely to support (or reject) both. If you explain the different genetic implications – that changes to the gametes can be passed on to future generations and so change the gene pool for the population as a whole – your students may wish to reconsider their views.

When selecting a context you may want to consider: the extent to which your students will be familiar with the context, the amount of information they will need in order to relate to it and the type and complexity of the issues that it raises. You will also want to consider the directions that the discussion might take and the additional issues that might be considered. There may be some issues that you would wish to avoid. For example, if you choose prenatal screening as the context, the discussion is likely to include a consideration of abortion of an affected fetus. If your class includes a student who suffers from a genetic disease you might want to avoid this context.

7.6 Genomics

Previous knowledge and experience

Students will need a good understanding of the ideas presented in Sections 7.1 and 7.3, including some understanding of what is meant by a mutation and how this can affect the gene product.

Genomics is a relatively new and rapidly developing discipline, and while the basic concepts illustrated in this section are unlikely to change you may need to look out for additional materials to ensure that your own knowledge is up to date and that you are aware of topical examples that might particularly interest your students. Good starting points for doing this are the Wellcome Trust (www. wellcome.ac.uk) and the Wellcome Trust Sanger Institute sites (www.yourgenome.org), and also www.insidedna.org.uk/.

A teaching sequence

As has been noted in earlier sections, relatively few single gene traits (characteristics or disorders) have been identified and while a linear view of genetics and gene expression (1 gene → 1 protein → 1 characteristic) is useful for developing an understanding of basic genetic principles, it doesn't tell the whole story. Genes often have multiple effects rather than just one (they are pleiotropic) and the majority of characteristics are determined by a complex interplay between multiple genes and their environment, over a period of time. These interactions can result in increased susceptibility to particular medical conditions in some individuals, probably as a result of increased susceptibility to environmental triggers such as alcohol, tobacco or asbestos. This section provides some accessible examples and explanations which may help your students to develop some understanding of this complexity.

At present this section is better suited to more advanced students, but within a few years genomics may become part of the general biology curriculum. For less advanced students it might be a good idea to focus on the main example, the development of cancer, to illustrate the key features of genomics: more than one gene is involved; there are environmental influences (which in this case result in mutations); it takes a combination of several environmental impacts (in this case mutations) to trigger a change (in this case loss of control of cell division leading to the growth of a tumour).

■ Cancer development

This example gives an indication of how multiple genes and the environment interact within an individual – in this case resulting in a tumour – and how bioinformatics (the analysis of genomic databases) can provide information which might be useful.

Tumours develop when a cell goes out of control and begins to grow and divide faster than neighbouring cells. Cell growth and division are normally controlled by two sets of genes – promoter genes, which promote growth and division; and suppressor genes, which suppress growth and division. If mutations (caused by environmental factors such as ultraviolet light and carcinogenic chemicals) occur in promoter genes, affected cells may start to grow and divide faster, i.e. it is the mutation that causes the gene to promote cell division. If the suppressor genes are functioning normally they will suppress this increased rate of growth and division and the cell will continue to function normally. If mutations also occur in the suppressor genes they may become inactivated and unable to suppress cell growth and division. If this happens the cell becomes cancerous, dividing rapidly to produce a tumour. It is thought that at least five mutations, across both sets of genes, are needed for a cell to become cancerous.

Because mutations are randomly occurring events, each cancer is caused by a unique combination of mutations within the cancer-related genes. By sequencing DNA from the cancer cells it is possible to identify the specific mutations which have caused the cancer in any individual patient. Eventually it may be possible to use this information to tailor the treatment of cancer to the individual patient. More advanced students can practise using real data to identify a set of these mutations using the 'Investigating Cancer' activity, available from www.yourgenome.org/teachers/kras/shtml* and described in *Science in School* (2010) Issue 16 p. 39 (www. scienceinschool.org).

Note: if you cannot access this website directly, try www. yourgenome.org/landing_teachers.shtml then select 'cancer, KRAS cancer mutation' from the options box at the side.

■ Epigenetics

The sequence of bases within the DNA determines the possible gene products, but nothing can actually be produced unless the gene is activated (switched on) and this is strongly influenced by environment. Epigenetics refers to the processes by which gene expression is controlled and the impact of this gene regulation on the phenotype.

One common mechanism by which genes are inactivated (switched off) is DNA methylation. After methylation, the sequence of bases in the DNA remains the same but its structure changes slightly, making it difficult for enzymes to bind. The effect this can have on the phenotype can be seen in Agouti mice, which produce brown, yellow or mottled fur depending on how the *Agouti* gene is methylated during embryonic growth. When pregnant mice are fed methyl-rich supplements such as folic acid and vitamin B12, most of their young have brown fur; when pregnant mice are not fed supplements, most of their young have yellow fur.

Some epigenetic events are irreversible and the affected genes are permanently switched off. There are ~25 000 genes in the human genome (estimates vary from 20 000–30 000) and in a young embryo each of these 25 000 genes has the potential to be active in every one of the unspecialised cells. As different combinations of these genes are inactivated in different cells, the cells become specialised in different ways (becoming, for example, nerve or muscle or bone cells). These particular epigenetic changes are irreversible and will be passed on, along with the genome, each time the specialised cell divides. In this way new cells will always have the same phenotype as the parent cell (the same structure and function). This epigenetic process is called selective silencing and results in the development of a new organism.

The extent to which epigenetic changes can be passed on to new somatic cells during mitosis has been surprising and there are many examples including control of flowering time in plants, linked to previous exposure to low temperatures, and mosaic patterns of gene expression in the phenotype linked to random inactivation of the second X chromosome at an early stage of development (see tortoiseshell colouring in cats in Section 7.2, enhancement ideas). There is even some evidence that epigenetic changes can be passed from parent to child (see www.epigenome.eu/en/1,63,0).

The extent to which epigenetic changes build up over a lifetime and so shape our phenotype – to an extent that can differentiate between genetically identical twins – has also been surprising (see www.epigenome.eu/en/1,4,0).

Stem cells

Knowing that genes, and even whole chromosomes, are selectively silenced within specialised cells (see Epigenetics, above) should help students to understand what stem cells are and why they are commercially and medically important.

Stem cells are cells that have the capacity to continue specialising. A small number of cell types within a mature organism retain some of this capacity, for example bone marrow cells, but the best (most useful) type of stem cell is found in the early embryo. At this stage none of the genes have been silenced and the cells still have their full capacity to specialise (they are totipotent). This means they have the potential to be used in cell therapy, to treat incurable illnesses and injuries caused by damaged or malfunctioning cells (Parkinson's disease or spinal cord injuries, for example). Using embryonic cells in this way clearly raises moral, ethical and legal issues relating to: the source of embryonic stem cells (e.g. early termination of pregnancy through miscarriage or abortion), ownership (who owns the resulting embryonic tissue) and informed consent (Are parents aware that the resulting embryonic tissue might be used in this way?); the point at which human life begins, the rights of an embryo (including the right to life) and the comparative rights of different individuals (the embryo, the mother, the father). There are also practical difficulties. Just as an organ transplant will be rejected if the differences in tissue type between donor and patient are too great, so it is with cells. Where mature cells are being used, as in bone marrow transplants, the whole population is potentially available in the search to find a good tissue match. As embryonic cells cannot be selected in this way, the risk of rejection is very high. These difficulties will need to be resolved before stem cell therapy can become a routine treatment.

 The Biochemical Society has a good resource on stem cells designed for school students (see: www.sciberbrain.org/Home/Stemcells.aspx).

■ Social and ethical issues

The extent to which a characteristic is determined by genes or environment, and the complexity of interactions between these factors, can have important social, political, ethical and moral implications. For example:

- If intelligence is determined mainly by genes, then the value of education for all may be questioned.
- If there are genes for criminal behaviour, then the responsibility of the criminal for their actions may be questioned.
- Conversely, if criminal behaviour can be mainly attributed to environmental conditions, then there may be a moral and social responsibility to change those conditions.
- If a local industry is based on a chemical that is known to be carcinogenic, and it has been shown that some people are more susceptible to this effect than others, how should the employer respond: say nothing; tell employees but leave decisions about

whether or not to be tested up to them; insist all potential employees are tested and then only employ the least susceptible; insist all current employees are tested and then sack those that are susceptible; offer employees a test and provide additional protection for any who find they are more susceptible?

Such issues could be discussed through consideration of a topical example (see Section 7.5 for guidance).

Other resources

Websites

It is sometimes difficult to keep up to date in a field that is changing as rapidly as genetics and genomics but a number of organisations provide good materials, freely available on the web, which are aimed at school students and their teachers. At the time of writing these include the following (but remember that websites do change):

www.abpischools.org.uk/: The Association of the British Pharmaceutical Industry website, with resources for students and teachers, including a module on genes and inheritance and an animation of meiosis.

www.ncbe.reading.ac.uk/NCBE/PROTOCOLS/menu.html: The National Centre for Biotechnology Education was the first biotechnology education centre for schools and specialises in how to make gene technology safe and accessible in the school classroom.

www.pbs.org: An American educational site that collates an impressive range of materials and resources from the net. Go to the following url and select the genetics topic: www.pbs.org/teachers/classroom/6-8/science-tech/resources/.

www.wellcome.ac.uk: The Wellcome Trust website, good for updating but see also the specific edition of *Big Picture* dedicated to genetics: www.wellcome.ac.uk/Education-resources/Teaching-and-education/Big-Picture/All-issues/Genes-Genomes-and-Health/index.htm.

www.yourgenome.org is the Wellcome Trust Sanger Institute site and www.yourgenome.org/landing_teachers.shtml provides an excellent resource for teachers.

www.insidedna.org.uk/: This website is designed to keep you up to date on the latest developments in the fast-moving world of human genomics and provide a forum for everyone to discuss and debate the issues raised.

www.geneticalliance.org.uk/: For information about genetic disorders.

www.scienceinschool.org: A journal written for school students and teachers drawing on cutting edge research.

Other useful websites

www.centreofthecell.org/

http://bio.edu.ee/models/en/: Cell World interactive for all cell processes.

www.ebi.ac.uk: for bioinformatics (DNA databases).

www.scizmic.net/: ideas for debates (including examples of how to organise students for such activities).

www.beep.ac.uk: includes information to support teachers in developing their own understanding of ethics and how to support ethical discussions in the classroom.

www.at-bristol.org.uk/cz/teachers/czhowto_web.pdf: managing discussion of controversial issues.

www.windfalldigital.com/ethicalemporium/: Ethical Emporium – for teachers; guidance on how to manage ethical discussions plus some examples.

www.sciberbrain.org/: a selection of teaching resources built around topical genetic contexts.

www.education.leeds.ac.uk/research/cssme/projects.php?project=88&page=1: Interactive teaching units developed for the National Strategy, see the unit on 'Key concepts in genetics'.

Background reading

Jones, A., McKim, A. and Reiss, M. (Eds) (2010). *Ethics In The Science And Technology Classroom: A New Approach to Teaching and Learning*. Rotterdam, Holland: Sense Publishers.

Banet, E. and Ayuso, G. E. (2003). Teaching of biological inheritance and evolution of living beings in secondary school. *International Journal of Science Education*, **25** (3), pp. 373–407.

Boerwinkel, D. J. and Waarlo, A. J. (2009). *Rethinking Science Curricula in the Genomics Era*. Full text available at http://igitur-archive.library.uu.nl/bio/2009-1208-200117/UUindex.html.

 Boerwinkel, D. J. and Waarlo, A. J. (in preparation, expected late 2011). *Genomics Education for Decision Making*. See: http://igitur-archive.library.uu.nl/bio/2011-0815-200626/UUindex.html

Corrigan, D., Dillon, J. and Gunstone, D. (Eds) (2007). *The Re-emergence of Values in the Science Curriculum*. Rotterdam, Holland: Sense Publishers.

Dolan, A. (1996). 'The making of Mandy'. *Journal of Biological Education*, **30** (2), pp. 94–96.

Ratcliffe, M. and Grace, M. (2003). *Science Education for Citizenship: Teaching socio-scientific issues*. Maidenhead: Open University Press.

Shaw, A. (1996). DNA makes RNA makes protein. *School Science Review*, **78** (283), pp. 103–105.

Shaw, A. (1997). Alien alleles. *School Science Review*, **78** (284), pp. 108–111.

Venville, G. and Donovan, J. (2008). How pupils use a model for abstract concepts in genetics. *Journal of Biological Education*, **43** (1), pp. 6–14.

8 Classification, variation adaptation and evolution

Neil Ingram

8.1 Classification
- Living organisms and non-living objects
- How many kingdoms?
- Plants and animals as major groups
- Classifying living organisms using similarities and differences
- What is a species?

8.2 Variation
- Variations occur between different species
- Variations occur within a species
- Continuous and discontinuous variation
- Variations may be inherited, environmental or (usually) both
- Nature and nurture in human development

8.3 Adaptation
- How are living organisms suited to their environment?
- Adaptations provide survival value
- Adaptations arise over time by evolution through natural selection

8.4 Evolution
- Ideas about evolution
- Timescales of evolution
- DNA and fossils as evidence for evolution
- Natural selection as the mechanism of evolution
- The birth and death of species
- Evolution and biodiversity
- Controversial evolution

Choosing a route

The four topics of this chapter do not fall into a single continuous teaching sequence, but will probably be visited and revisited several times across the primary and secondary science curriculum. Table 8.1 shows one possible model for a teaching scheme that starts with the adaptations that organisms show to living in specific habitats (10–11 years). The possibility of introducing evolutionary teaching into primary education is being actively considered in England and Wales.

Variation and classification are introduced later (12–14 years) with evolution introduced for those between 14–16 years. It is likely that the interdependence of organisms will also be introduced at this stage and this lends itself to a consideration of the extinction of species in terms of its impact on evolution and on biodiversity.

Table 8.1 Model for a teaching scheme

Stage of schooling	Teaching sequence
Late Primary (10–11 years)	about the different plants and animals found in different habitats; how animals and plants in two different habitats are suited to their environment
Early Secondary (12–14 years)	all living things show variation, can be classified and are interdependent, interacting with each other and their environment
Intermediate Secondary (14–16 years)	organisms are interdependent and adapted to their environments; variation within species can lead to evolutionary changes, and similarities and differences between species can be measured and classified
Note that in Scotland there are differences to this model.	

Evolution is a core scientific concept, and in those curricula that have differentiated pathways of progression (such as the GCSE Science and Additional Sciences courses in England and Wales) it may be introduced in the early stages of the course. It is not atypical in these schemes for students to be learning about evolution when they are 13 or 14 years of age. Evolution is one of the most important unifying ideas in modern biology, and teachers who use teaching schemes where evolution is taught early may need to find ways to keep revisiting this idea throughout the remainder of the course. This should not be too difficult as evolution has links to every aspect of biology.

8.1 Classification

It is possible to deal with the ideas of evolution from a number of different starting points, for example from the characteristics of living things (Chapter 1) or ecology (Chapter 9). You may decide to make the topic almost self-contained, dealing with the structure of the classificatory system at an appropriate level and how living things fit into it, introducing students to the features used for classifying the various groups, such as the vertebrates or the arthropods. A perfectly sensible alternative approach is one that delays formalising classification in this way until experience of working with a wide range of living things, e.g. in a local habitat, has been used to introduce some of their characteristic features. Whichever sequence you adopt it is likely that students will meet the ideas in the following order: classifying into living organisms and non-living objects; into plants and animals; and then into the major taxonomic groups of both. Some teachers may wish to introduce students to ideas about kingdoms at this stage.

Previous knowledge and experience

Students will probably have done some work on organisms that occur in their local environments. They will have experience of the characteristic features of living things: that they feed, move, grow and reproduce. They will have made collections of plants and animals and used various identification aids, including keys, to classify them. Students will vary considerably in their confidence and competence with these ideas.

A teaching sequence

■ Living and non-living

At primary level/early in secondary school, students identify the seven characteristic features shown by all living organisms, often using a mnemonic such as **Mrs Gren** (**m**ovement, **r**espiration, **s**ensitivity, **g**rowth, **r**eproduction, **e**xcretion and **n**utrition). At this stage it is unrealistic to expect students to have more than a sketchy idea of each of these life processes. That will be, in essence, the subject of their entire biology course.

A dead organism is one that used to do these seven processes, but no longer does so. Viruses are microorganisms that are unusual because whilst they can reproduce, they lack, on their own, any of the other six life processes, since they invade the cells of a host organism and use those cells to complete their life cycles. For these reasons, many biologists refuse to accept that viruses are living organisms.

One possible teaching approach is to examine collections of living organisms and non-living objects. Where possible the displays should be of the actual objects and organisms, perhaps arranged into a 'circus' of exhibits that can be supplemented by a variety of pictures and video clips. Students can rate the exhibits as 'living', 'once alive' or 'never alive'. It is important to give students the time and opportunity to discuss the reasons for their classifications with each other. The display should include some less straightforward examples such as an apple fruit, a seed, a piece of coral, a fossil, a piece of rock, a skull, a bone, a labelled picture of the HIV virus or a picture of a fetus in the uterus. Students should come to appreciate that the identification of an object as living is based on it displaying all of the characteristics at some point in its life history, although not necessarily at the same time. Further investigation may be needed to reveal evidence that some of the processes of life are occurring.

Research and everyday experience indicate that some young children think of the term 'animal' as being the same as 'mammal', even though they can successfully recognise other animals. Thus, a fly is correctly identified as an insect, but not considered an animal. Humans are humans and may be viewed neither as mammals nor as animals (based on Driver *et al.*, 1994). These are misconceptions that indicate that some students do not appreciate the hierarchy of the classification system.

■ How many kingdoms?

Traditionally, the highest level of the classification system proposed by Linnaeus in the eighteenth century was the kingdom. The levels of the hierarchy beneath this level are, in descending sequence: phylum, class, order, family, genus and species. Asking students to form a mnemonic is a good way of reinforcing this sequence. The simplified full Linnaean classification of our species is:

- Kingdom: Animal
- *Phylum: Vertebrate
- Class: Mammal
- Order: Primates
- Family: Hominidae
- Genus: *Homo*
- Species: *Homo sapiens*

The full classification gives not only the name of the organism, but also its precise location in the living world. It is, if you like, its name and its address. It can be helpful to get students to find other examples of species within each of the groups and to discover the characteristics that all organisms in that category share in common. Biologists call the categories 'taxa' (singular, 'taxon').

Classifying the entire living world in this way has proven to be very difficult and controversial. Scientists find it difficult to agree on what might appear to be even the simplest of issues, such as how many kingdoms there are. At different times in history there have been two, three, four, five or even six kingdoms.

The five kingdom model is probably the most useful at this level. It consists of: the Monera (bacteria), Protoctists (single-celled organisms and some seaweeds), Plants, Fungi and Animals.

Students who go on to advanced studies may encounter a six kingdom model. This model seems to reflect more reliably current

*Technically, the group Vertebrate is a sub-phylum of a larger group called the Chordates, which includes animals with primitive backbones. For many students the simplified version will be sufficient. The name of the species is usually underlined when written by hand or printed in an italicised font.

knowledge about evolution. The model creates a higher category than Kingdom, called a Domain. There are three domains: Archaea (primitive bacteria), Bacteria and Eukaryotes. (Archaea and Bacteria are often combined into a single group called 'Prokaryotes'.)

The Eukaryote domain contains the Protoctists, Plant, Fungi and Animal kingdoms. The key teaching point is that there is no single 'correct' classification system, and students should be taught the one that is most useful to them depending on their age and other factors, whilst being made aware that other systems are sometimes used for specific purposes.

■ Plants and animals as major groups

There is only one characteristic that is shared by all green plants, that is the presence of chlorophyll. This means that green plants make their own food in photosynthesis. (Some parasitic plants lack chlorophyll, but this is thought to be an evolutionary adaptation that facilitates a parasitic lifestyle. They will have evolved from ancestors that possessed chlorophyll.) Animals do not possess chlorophyll and feed on plants or other living organisms.

There are differences in the structures of plant and animal cells. Plant cells have a cell wall and a permanent fluid-filled space called a vacuole. Animal cells have neither a cell wall nor a permanent vacuole.

A detailed account of the diversity of the animal and plant kingdoms is beyond the scope of this chapter, but students ought perhaps to be aware of two major groups of animals, the vertebrates (fish, amphibians, reptiles, birds and mammals) and the arthropods (including crustaceans and insects).

Only conifers and flowering plants reproduce using seeds. Conifers reproduce with cones and flowering plants produce flowers. A circus of living plant material (including ferns, mosses and liverworts) and algae can reinforce the diversity of plant and plant-like forms.

Access to living and museum specimens as well as video images can be provided. The Arkive website (www.arkive.org/) is an excellent place to visit for such images. Sorting, identifying and then classifying are much more interesting when a motivating activity is being undertaken. For example, students could sort through a collection of leaf litter either *in situ* in a wood or in samples brought into the laboratory. One large bag can contain sufficient leaf litter for a whole class. Students could carry out a similar exercise using pond water samples to extend the range of organisms involved. This can be extended by fieldwork in the school grounds, local park, on the beach or other suitable habitat (see Chapter 9).

These activities will allow students to use a wide range of identification aids, including keys. Younger students may need to be taught how to use identification keys, which can become complex through technical language. Online systems (such as the RSPB's bird identifier, www.rspb.org.uk/wildlife/birdidentifier/) can be an effective intermediate stage.

Further activities

- Sorting and grouping collections of pictures of animals and plants.
- Matching exercises, e.g. put in one group all the plants that have roots, stems and leaves and into another the plants that do not have these features.
- Crosswords based on matching names to descriptions.
- Concept circles, e.g. draw a circle and put in it the names of all the animals that have jointed legs, a hard outer skeleton and compound eyes. These are arthropods. Inside this circle draw another in which you put the names of all the animals that have three pairs of legs. These are insects.

Enhancement ideas

- Use databases to record, identify, group and sort animals and plants.
- Find out the characteristic features of different groups from a website (e.g. The Tree of Life: http://tolweb.org/tree/).

■ Classifying living organisms using similarities and differences

Classification relies on identifying similarities and differences between organisms. Many younger students may find it easier to notice differences than similarities; hence some encouragement to notice similarities may be needed. Describing two distinct types of woodlice, for example, without using any technical names of parts can lead to some interesting creative writing.

Students' prior experiences in classifying things (rocks, chemicals, materials, etc.) may have involved using a single characteristic (such as: Does the material conduct electricity or not?). Biological classification is more complex. Taxonomy, the principles on which this classification is based, involves the use of multiple characteristics, the groupings are hierarchical and they have an evolutionary significance. Organisms close to each other in the classification (such as species within the same genus) share recent common ancestors, whilst those further apart (such as species linked at the level of the phylum) shared a last common ancestor in the more remote past.

■ What is a species?

The fundamental unit of the Linnean system of classification is a species. The definition of what constitutes a species is rather loose and has give rise to debate and controversy over the years. Recent advances in DNA technology that allow us to compare the genomes of related organisms have often served to muddy the waters still further.

The academic controversy surrounding the 'species concept' has been intense, with up to five different definitions being used at certain times. Central to the idea is that of morphology (the shape and visible structure of the organism): individuals of a species have one or more characteristics in common that enable them to be recognised as a group. These characteristics are not shared with other related species, so the species look distinctive.

Two species of oak tree in Europe illustrate this point clearly. The pedunculate oak (*Quercus robur*) has leaves that are attached to the branches by a very short stalk 3–8 mm long. In contrast, the leaf stalks on the sessile oak (*Quercus petraea*) are much longer (1–2 cm). The diagrams below illustrate the leaves. These are shown in Figure 8.1.

Figure 8.1 The distinctive leaves of two species of European oak tree

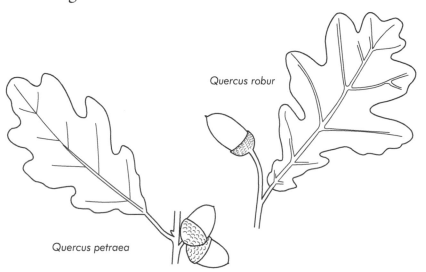

Quercus robur

Quercus petraea

These species are easy to identify and have traditionally been found in different habitats. They are good 'morphological species'. However, when they grow close to one another, they interbreed to produce hybrid offspring that are fertile. These offspring have characteristics that are intermediate in appearance between the two parental types. Over time, these two distinctive species will gradually blend into a single species.

This brings us to the second definition of species. The individuals within a species are (in principle) able to breed to produce offspring that are fertile. This is the definition of a biological species.

The two oak species above are good morphological species, but not ideal biological species. Are they one species or two? This is a matter of debate. Whilst they live in separate and distinct habitats then they exist as distinct taxa. Their future as two species depends upon whether the hybrid forms become established, reproducing with each other and with the original species. The idea that species can form and disappear is something that can be revisited in teaching about evolution, especially at higher and advanced levels.

In the animal world, a horse and a donkey are good morphological species, but can interbreed to produce a mule. Horses and donkeys are only classified as separate species because the mule is sterile and cannot breed further.

Some good biological species are hard to identify because they look so similar. For example, the western meadowlark, *Sturnella magna* and the eastern meadowlark, *Sternella neglecta* are two distinct species. The species look alike and co-exist in the same habitats. Photographs of the two species side-by-side can be seen at the web address http://www.nature.com/scitable/topicpage/why-should-we-care-about-species-4277923. The western meadowlark and the eastern meadowlark have distinctly different songs. They do not interbreed and are classified as separate species.

Hey has discussed the implications of the species problem further (Hey, 2009). The practical investigation of species will be developed in the next section.

8.2 Variation

Three main strands are covered in studying variation:

- examples of variations between and within species (interspecific and intraspecific variation)
- origin of variations and their possible inheritance
- adaptive and evolutionary implications of variation.

This section covers the first bullet point, origins and inheritance is mainly covered in Chapter 7 and evolution later in this chapter.

Large numbers of examples of intraspecific variation occur in other topics in biology. Students will meet the idea in considering the breathing, heart and pulse rates of members of the class. In growth experiments that use plant cuttings, seedlings, duckweed, etc., although other teaching objectives will be pursued, the chance to reinforce the idea of variation can be taken. Eliminating variation between different batches of experimental material is often considered as an important aspect of the design of investigations. Variation is universal in plant and animal species and is the raw material of evolution. This idea is best developed by studying living material at first hand.

Previous knowledge and experience

When classifying living things, the idea that variation occurs between different 'sorts' of organisms will have been used. This work will have concentrated on interspecific variation and builds upon the species concept developed earlier.

A teaching sequence

■ Variation between species

It is useful to illustrate that organisms that may look quite similar can belong to different species or even to different genera. Water boatmen (two common genera, *Notonecta* and *Corixa*) and woodlice (examples of four common clearly distinguishable genera are easily found: *Oniscus*, *Armadillidium*, *Porcellio* and *Philoscia*) are useful for this work. Specimens of some of the common species of buttercup can also be used (picked, not uprooted). In these examples, the names of the organisms are not important; indeed the different specimens could be given reference letters or numbers instead of names. Students should be encouraged to observe and record the different features that could be used for separating the species into different groups (taxa). A difficult idea, which nevertheless should be tackled, is that some characters are of use in classifying but others are not. Students could be introduced briefly to the developmental variation that can occur, most obviously in sizes at different ages, but also in form and structure, e.g. the different stages in insect and frog life histories.

■ Variation within species

A useful starting point is to have students consider some examples of the results of artificial selection. There are a number of different possibilities and it is not essential that all students consider the same material. Pictures of different breeds of dogs or domestic fowls are easily collected. Seed catalogues are also useful to provide examples and descriptions of different varieties of flowers and vegetables, e.g. potatoes, radishes, broad beans, peas, dahlias and roses. Students should identify the different characteristics that have been selectively bred and the possible reasons for this (not methods of doing so at this stage). Some of this information, e.g. features of different varieties of potato, could be entered into a spreadsheet. Ideas worth considering are the shapes, sizes and colours of the tubers, how growth rate varies in first and second 'earlies' and 'main' crop varieties, planting and harvesting recommendations, different yields, the different nature of varieties that are recommended for boiling, roasting and making chips, eelworm resistance, and so on.

■ Continuous and discontinuous variation

It is usual to distinguish between continuous and discontinuous variation. In the former the characteristic concerned does not fall into discrete categories; in the latter it does and any individual either possesses or does not possess that characteristic or trait. This distinction may be introduced by taking some interesting human examples.

Such examples may be taken from the class itself, always being careful to avoid potential embarrassment of students who may be sensitive about differences they may possess. Ear lobes can be unattached (dangly) or attached. The irises of the eyes may be brown, blue or green. Blood groups can be A, B, AB or O. Hair can be blond, brunette, red, or black.

The division of variation into continuous and discontinuous is rather artificial, as a consideration of the eye colour example shows. There are a range of blue eyes and a range of brown eyes and some colours (e.g. hazel) may be harder to classify. Other examples of discontinuous variation are: normal haemoglobin and haemoglobin S (the sickling trait), the ability to taste phenyl thiocarbamide (PTC), and fingerprints (divided into the four basic types). Some of these examples (such as fingerprints) lend themselves to some practical work by the students. Carry out a risk assessment if you want to use PTC in a practical activity.

Appropriate human examples of continuous variation are: height and weight, blood pressure, heart rate and cholesterol levels.

There are many opportunities to study variation in other organisms.

Buttercup (*Ranunculus*) flowers are a good source of natural variation. Collect a large number of flowers of the same species of buttercup or lesser celandine. Distribute them among the class and ask them to find out if all the flowers have the same number of petals. Several hundred flowers can be examined in a few minutes by a class. This will lead to the collection of data, from which a bar chart can be drawn. Students will find a wide variation in the number of petals, five probably being the modal value, depending on the species, with the possible range being from four to 20. This is an example of a variation where the character involved, the number of petals, is a whole number. It is because of this, of course, that the data are plotted as a bar chart rather than as a histogram because no class intervals are selected. It is important that students learn how to present such data appropriately.

The lesser celandine (*Ranunculus ficaria)* can be used for a number of activities. The plant is widely distributed, has a long flowering time and flowers at a 'convenient' time, early spring, for school work.

Samples of the leaves are taken from different plants (or, even better, from populations growing in different places). Students measure the length and breadth of the leaves, plot these data as scatter graphs and compare differences.

A portion of the lower epidermis of the leaf can easily be stripped off after tearing the leaf. This is mounted in a drop of water on a microscope slide. Under high power, chloroplasts can be seen clearly enough in the guard cells to be counted. Samples of about 20 guard cells taken from different leaves can show wide variation in chloroplast numbers.

Fruit flies: Students can sort through a collection of freshly killed or preserved (in 50% ethanol) fruit flies that comprise wild type flies and examples of visible mutations, such as vestigial wing, curled wing, white eye and ebony body.

Primroses: In spring a small bagful of primrose flowers can be collected in a few minutes. But be aware of any legislation about this. Students can 'dissect' the flower and discover the different arrangement of stamens and stigmas constituting pin and thrum flowers (see Figure 8.2). Some students might be able to deduce that these structures favour cross-pollination by insects.

Figure 8.2 Pin (left) and thrum (right) flowers of *Primula*

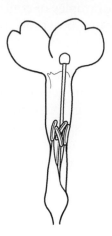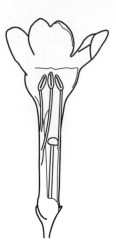

At this point in the teaching sequence you should discuss the idea of inherited or environmental origins of variation. It will be inevitable in studying some of these examples that ideas about the significance of the variations will emerge. The extent to which variations are considered to be adaptive will vary from example to example. There is an argument for presenting students with some examples of variations to which no obvious adaptive or selective significance can be attached.

Further activities

- **The brown-lipped, banded snail** (*Cepaea nemoralis*). The extensive polymorphism of this animal is well known. Snails differ in the background shell colour – pink, yellow or brown – and in the number and distribution of the prominent dark brown bands that occur. The number varies from 0 to 5. Differences in the width of the bands as well as their positions allow grouping into a very large number of categories (e.g. if a snail has a single band it can occur in the 1, 2, 3, 4 or 5 position, on a background that is either pink, yellow or brown). Collections of empty shells can be made from a variety of habitats. They keep well in screw-top jars and will retain their colour for years, especially if given a coat of nail varnish. This work can be greatly extended by comparing the frequency of various types found in different habitats (see later). Living snails may also be collected. They are easily maintained in the laboratory. They breed very easily and there is the possibility of using them for genetics experiments. Evolution MegaLab has a useful survey website on *Cepaea*.

■ Variations may be inherited, environmental or (usually) both

Some characteristics are inherited, others are environmental, yet most involve a combination of the two. Often those characteristics that are inherited have complex patterns of inheritance, particularly in humans, though blood grouping is a fairly straightforward example of inherited characteristics. The ABO blood group system is well researched and the conclusion that there is a single gene with three potential alleles (A, B, O, with A and B co-dominant and O recessive to A and B) is well established. Many other 'textbook' examples of simple (Mendelian) inheritance in humans are more problematic. Ear lobe shape is often presented as being the result of a single gene with the allele for attached ears being recessive to that for unattached ears. However, this is not well established, and students should be discouraged from thinking it applies to all situations (The bright hub website discusses the complications, www.brighthub. com/science/genetics/articles/41418.aspx). Reiss (1988) has discussed the complexities of the inheritance of human eye colour, where brown-eyed children may have blue-eyed parents. Tongue rolling is sometimes presented as a genetic trait, but there is evidence that it can be developed by learning (see http://udel.edu/~mcdonald/ mythtongueroll.html for a critique of the standard account).

One way into this area of work is by considering human family trees. A wide selection is available to draw on, e.g. the royal haemophilia, the Habsburgs' lip and the occurrence of marked scientific ability in the family to which Josiah Wedgwood, Charles Darwin and Francis Galton belonged.

Students will be aware, with a little prompting, of some interesting effects of the environment on humans. Sun-tanning, dieting, the effects of weight training and other health-related activities will be familiar examples. Changes to the body caused by such activities are caused exclusively by the environment and cannot be passed on to future generations.

Most biological characteristics are controlled by an interaction between genetic and environmental factors. Continuous variables, such as height, will be controlled by a large number of genes exerting individually small effects. The environment will also influence the expression of these genes. The equation:

$$\text{phenotype} = \text{genotype} + \text{environment}$$

is a useful introduction to thinking about the expression of biological characteristics. Phenotype describes the characteristic being observed and genotype describes the genes. Genes interact with the environment to produce the phenotype.

A convincing demonstration of the interaction between genetic and environmental factors is available using chlorophyll-deficient seedlings. In tobacco a recessive allele for chlorophyll deficiency results in albino seedlings that survive for a few days only. Seeds from F1 heterozygotes (from Philip Harris) are sown on black agar in Petri dishes. After 10 days the seedlings produce cotyledons and the plants, when scored, will be found to occur in the ratio
3 green : 1 albino. If they have covered work on genetic ratios, students will see that the results provide evidence that cotyledon colour is an inherited characteristic. The students are then presented with further seedlings that were sown at the same time as the first batch but were subsequently maintained in the dark (e.g. in a cupboard). All the seedlings will be seen to be albino. These seedlings, when placed in the light for several days, will produce a pattern of results similar to the earlier ones, i.e. 3 green : 1 albino. This illustrates that an allele for chlorophyll production has to be present to produce green cotyledons but that its expression depends on the presence of light.

■ Nature and nurture in human development

Traditionally, human characteristics were thought to have been caused by inherited factors (nature) or by the effects of the environment (nurture). This was, largely, a philosophical debate that raged in biology and psychology and had social impacts in areas such as education (selective schools versus inclusive schooling) and racial policy (where some 'races' were deemed genetically 'inferior' to others). The dichotomy has been shown to be false, even down at the level of the nucleus, where DNA (the source of all genetic information) is entirely dependent upon its chemical environment (enzymes etc.) in order to function. The reality is that it is not 'nature or nurture' as much as 'nature via nurture' (Ridley, 2003).

Recent advances in genome studies have yielded a lot of information on the genetic basis of many human diseases and conditions, such as breast cancer, diabetes, schizophrenia, bipolar disorder and alcoholism. In each of these conditions, genetic factors interact with the environment. Genes may predispose a person to get a particular condition, but so can certain environments. The people at the highest level of risk are those where the effects of genes and the effects of the environment reinforce each other. For example, young teenagers who smoke cannabis are ten times more likely to develop schizophrenia if they have two copies of a faulty version of the COMT gene than if they have two copies of the normal form of COMT (Lawson, 2005).

It is now possible to survey whole genomes, looking for variations in single bases of DNA (these are called single nucleotide polymorphisms, SNPs – pronounced 'snips'). This is revolutionising our understanding of disease and could eventually lead to personalised medicines where drugs are formulated appropriate to each person's genotype. An understanding of genes and the environment will impact on the choices that people can make throughout their lives.

8.3 Adaptation

Previous knowledge and experience

Students will have met the idea of adaptation in the context of animals and plants being 'adapted to their environments', e.g. fish being adapted to swim in water and plants adapted to harvest light.

A teaching sequence

Here are some examples of the use of the word 'adaptation' in biology:

- People visiting high altitudes for several weeks adapt to the lower oxygen content in the air by producing more red blood cells.
- On entering a dark room, human eyes take a few minutes to adapt to the lower light intensity there.
- Mammals are adapted to live on land.
- The leaves of desert plants are adapted to reduce water loss by evaporation.

These examples show that two quite different meanings are given to adaptation: the first pair refers to a process that involves a change of some sort taking place in the short term and the second pair describes the outcome of a very long-term series of changes.

There is another difference between these statements. Adaptation to lower oxygen content or adaptation of the eye when someone enters a dark room involve the individual; adaptation to life on land takes place only over many generations, and such adaptations are principally the result of the process of natural selection. Natural selection has nothing to do with the short-term changes apart from having brought about the evolution, over time, of systems that can make these types of short-term physiological adjustments.

Students can become extremely confused about the different meanings that they hear attached to the word 'adaptation'. Their confusion is potentially dangerous because inability to distinguish between these nuances of meaning leads to misconceptions. Students should be alerted to the different ways in which the term is used.

When is a particular feature of a living organism actually an adaptation? It is unscientific to assume that because an organism living in a particular environment shows some structural feature that is different to that of another organism in a different environment the feature must be an adaptation. Unless scientific evidence is available to measure the benefits of the 'adaptation', we can only hypothesise that a feature is adaptive.

All adaptations are relative. If an organism is adapted to a particular environment, it must also be less well adapted to a different one. It is also often better adapted to that environment than is another organism that does not live there. Sometimes an invader in a new habitat might be better adapted than the residents and replace them. Grey squirrels have replaced red squirrels over most of the UK, for example. These types of thinking are useful because they illustrate the need for experimentation and suggest some of the different experimental designs that may be involved.

■ Teleology

The word 'adaptation' also crops up in everyday language, for example:

- Some Jane Austen novels have been adapted for television.
- An old chimney pot can be adapted for use as a plant container.

In both of these examples there is a sense of purpose involved, a deliberate intention to make something to achieve an objective. When similar statements are used about aspects of biology they are said to be teleological. Such statements are very common, particularly when structure and function relationships are involved. They usually include the words 'design' or 'need' or the phrase 'in order to'. Here are some examples:

- The coat of the mountain hare turns white in winter in order for the animal to be camouflaged in snow-covered country.
- The leaf of a plant is designed to bring about efficient gaseous exchange.
- The walls of the alveoli need to be thin so that diffusion of gases is efficient.
- Arteries have thick walls in order to be able to carry blood at high pressure.

Many biology teachers not only see nothing wrong with using expressions of this sort, but also deliberately use teaching techniques based on such thinking: 'Design an animal that lives in ... and feeds on ... Say why you gave it the features you did'.

However, some of these statements, like the first one above, are potentially misleading in that they suggest that the organism showed some purpose in developing the adaptation so that it could achieve the intended outcome.

One objection to this teaching approach is that it might imply that if something is designed there must be a literal designer outside the system. As Richard Dawkins (1986) has famously argued in *The Blind Watchmaker*, there is no such designer; the watchmaker is natural selection and that is blind; it has no end in sight.

■ Anthropomorphisms

It is easy for students to say that microbes 'learn to become resistant to modern antibiotics', that the hare 'knows that winter is approaching and grows a white coat' or that a virus 'tricks the cell into copying its genetic code'. Each of these is an example of anthropomorphism, where human emotions, needs or competencies are attributed to other living things. Whilst such expressions may be used for popular effect, perhaps they are best avoided in your students' writing, since they can lead to the idea of intentionality and teleology. Or if they are used, students should demonstrate that they realise they are metaphors. Ideally, students should develop the ability to write, at times, without such anthropomorphisms ('hares that have an inherited tendency to grow a white coat at the approach of winter are less likely to be eaten and so leave more descendants, leading to the spread of the trait through the population').

■ Teaching adaptation

Opportunities to consider adaptation will occur throughout teaching schemes in the 11–16 range. In ecology work you can use the ideas when considering the unequal distribution of organisms in a habitat. When comparing the different living things found in, say, running and standing water there will be a range of ideas to develop. The effect of abiotic (physical) factors on distributions is likely to involve a consideration of adaptation. Interactions between organisms, particularly when considering 'what eats what', will provide additional ideas. Most topics involving flowering plants and mammals will allow discussion of examples of adaptations. There are large numbers of potential examples in other chapters of this book. The points mentioned earlier can be drawn on to inform a teaching approach.

8.4 Evolution

Coverage of this topic may occur in the last year of compulsory secondary education. Here students will be able to draw on the wide range of ideas they will have gained from their work on variation, adaptation and genetics. On the other hand, in many English and Welsh GCSE courses it now appears as part of a common core and features in Year 10 (14–15 years). It is important, in such cases, that evolution is revisited often in later parts of the course. It is a principal idea in biology that underpins every aspect of the subject.

■ Ideas about evolution

Timescales of evolution

The Earth is about 4.5 billion years old (4.5×10^9 years). It is possible that some of the simplest chemicals needed for building living organisms were formed in the seas. Some may have come from outer space. The development of replicating molecules (such as DNA) and proteins were a precursor to the formation of living cells.

Theobald (2010) has shown that all life on Earth appears to have a single common ancestor that probably lived on Earth about 3.5 billion years ago. This could explain why the DNA code is the same in all living organisms. The earliest living organisms were probably similar to today's Archaea bacteria, such as methanogens that metabolise hydrogen and carbon dioxide into methane, halophiles that grow in salty environments and thermoacidophiles that thrive in acid and high temperatures (up to 110 °C). These organisms show adaptations that would have allowed them to survive in the primordial conditions on Earth. Certain prokaryotic bacteria may have formed a symbiotic relationship to produce the first eukaryotic cells. There are many similarities in the structure of mitochondria and chloroplasts, and free-living prokaryotic cells. Eukaryotic cells probably evolved about 2 million years ago.

Table 8.2 shows an approximate timeline for the evolution of a few important groups of living organisms.

The timescale involved is so vast that it is difficult for students to comprehend. BBC's *Life on Earth* envisaged the whole of life on Earth compressed into a single year, starting on January 1st. On that scale the first eukaryotes appeared in September of that year. Vertebrates appeared in mid November and modern humans appeared in the evening of December 31st of that year. Humans are the 'Johnny come latelies' of evolution.

Table 8.2 Approximate timeline for evolution

Organism	Age at which group appeared (years ago)
bacteria	3.8 billion
insects	400 000 000
vertebrates	525 000 000
mammals	200 000 000
flowering plants	130 000 000
humans	195 000

DNA and fossils as evidence for evolution

The traditional main evidence for evolution has been fossils. Fossils are the preserved remains of organisms from the past, usually imprinted in rocks. Sometimes whole insects can be preserved in amber – sticky resin from certain trees that hardens into a solid. The fossil record is notoriously incomplete, partly because relatively few organisms were preserved as fossils. Those fossils that are collected are often incomplete representations of the original organisms.

It is possible to compare their structure with living relatives and this 'comparative anatomy' has been useful in the past. Comparative anatomy can allow living specimens to be compared with each other. It has shown, for example, that the duck-billed platypus is a mammal, despite having characteristics of birds, reptiles and mammals.

The ability to sequence and compare the genomes of living organisms is transforming our understanding of evolution. We have within our cells DNA that provides a record of our evolution. We are starting to develop the ability to read and interpret the DNA record within. These analyses are throwing up some interesting surprises.

Modern humans, for example, are known to have evolved in Africa. We now believe that a small group of humans left Africa about 70 000 years ago, probably crossing the Red Sea into Southern Arabia. Genetic studies show that this group spread out to populate the whole of the rest of the Earth. All of the races of Europe, Asia, Australia and the Americas descended from this small group of pioneering humans. The genetic diversity within groups of Africans descended from those who stayed is far greater than the genetic diversity of any of the other populations of humans on Earth.

Natural selection as a mechanism for evolution

Students should examine evidence that some species of plants and animals died out in the distant past and that modern species are their descendants. This will usually take the form of considering aspects of the fossil evidence and how they support the theory of evolution. Most courses require a treatment of the mechanism of evolution by natural selection. This gives you a good opportunity to consider with students the nature of evidence and the role of theory in sciences in the context of what is the most important integrative idea in biology. You can also introduce students to the work of one of the most famous scientists of all time, Charles Darwin.

Darwin's ideas can be expressed simply:

- Members of a species are different from each other (show variation).
- Much of this variation can be inherited.
- Organisms compete with each other for resources in short supply.
- Some organisms have features that give them a better chance of surviving and reproducing.
- They pass on these features through their genes to their offspring.

This is evolution by natural selection, the 'survival of the fittest'. (Fitness refers to Darwinian fitness, not (necessarily) the most athletic.)

The birth and death of species

Most of the species that have ever existed are extinct. Evolution can allow species to form, and allow species to become extinct. Under certain conditions, species can form very quickly. We know, for example, that there are up to 500 species of cichlid fish in Lake Malawi in East Africa, all of which are less than 2 million years old. They show an extraordinary range of colours and forms, and are the staple of many tropical fish collections. They are morphological species, and survive because of behavioural mechanisms that ensure that each species only mates with its own kind. These mechanisms include visual clues (such as bright colours), unique nest-building behaviours by males that only trigger mating behaviour with females of that species and 'mouthbrooding'. This is where female fish keep the baby offspring safe in their mouths to prevent them from being eaten by predators. As adults, these fish tend to stay in the same locality. This results in different species of fish being restricted to their particular neighbourhood in the lake.

We now know that evolution has not proceeded at a constant rate throughout history. There have been times of rapid evolution (explosions) and also times of mass extinctions. For example, the fossils of the Burgess Shale record evidence of an explosion of evolution in the Cambrian period about 500 million years ago. Stephen Jay Gould (1989) wrote about these fossils in his book *Wonderful Life*. Many of the forms were so unusual that they had no modern equivalents, causing Gould to speculate that they must have become extinct in some mass extinction event. This led him to his celebrated view that if conditions on Earth had been different at any time, then evolution might have proceeded along a different path and humans would not necessarily have evolved. Simon Conway Morris, one of the scientists who worked with the Burgess Shale fossils, disagrees. In *The Crucible of Creation* (1998), he challenges Gould on the significance of chance in the evolution of advanced forms of life.

Evolution and biodiversity

It is thought that there have been five mass extinction events in the history of life on Earth. One of them, about 65 million years ago, included the demise of the dinosaurs. It took vertebrates 30 million years to recover from the effects of that extinction.

It is clear that we are experiencing a sixth mass extinction, largely as a result of human activity. It is estimated that the loss of species is as great as in other previous extinctions, largely caused by deforestation and other human activity. Climate change is likely to impact further on this extinction.

Linking the teaching of evolution to biodiversity adds a sense of perspective and urgency to our need to consider effective ways of conservation. Howarth and Ingram (2010) discuss this further.

Previous knowledge and experience

Many students are keenly interested in and knowledgeable about fossils in general and dinosaurs in particular. They will also bring to the class their own ideas about evolution (Deadman and Kelly, 1978). Some students' views about the mechanism of evolution will be rooted firmly in Lamarckian ideas: features acquired during the life history of the organism can be transmitted to the next generation. Research studies have revealed that these ideas may be developed long before students are formally taught this topic and are resistant to replacement by Darwinian interpretations. They may survive secondary school teaching and even early undergraduate teaching. Such thinking requires consistent and sensitive challenge (Brumby, 1979 and Engel, Clough and Wood-Robinson, 1985).

A teaching sequence

A possible teaching sequence at an introductory level is as follows:

1 Outline the main ideas involved in the theory of organic evolution: animals and plants alive today are descended from simpler ancestors. Point out that the theory has enormous explanatory power in accounting for the history of life on Earth.
2 Present evidence to support that theory. This will mainly be based on the fossil record. Extension work for able students might take in the evidence from geographical distribution, classification genomics and comparative anatomy.

3 Look at some examples of selection in action.

4 Explain how natural selection provides a mechanism for evolution.

■ Studying fossils

Ideas about extinction and the fact and rate of change are particularly important. Evidence for both can be shown by examining the fossil record.

- Examine a collection of fossils (real, models and pictures) that includes examples of organisms very similar to surviving ones (e.g. sea urchins, *Ginkgo biloba*) and very different from today's (e.g. *Archaeopteryx*). The well-known horse fossils probably provide the best example of a sequence of change over time (see Figure 8.3). Students should be told that the sequence presented to them is a simplified version of the record.
- Construct, to scale, a timeline (using string or a roll of paper) or a clock face that gives the history of the Earth in geological time. Get the students to place on the diagram the periods when the various organisms first appear in the fossil record. Work with major taxa of plants and animals. Talk about the significance of the sequence of increasing complexity that is shown.

■ Selection

Selection provides part of the explanation for the mechanism of evolution (variation and its origins is another part). It is important that students understand that natural selection is an ongoing, widespread and rapid phenomenon that occurs in the present day, as well as having shaped past events. It is also important to stress that selection does not automatically lead to speciation (the formation of new species). This point is dealt with below.

■ A simulation

A useful practical starting point is to do a simulation of selection using red and green cocktail sticks (plastic or dyed wooden ones). A hundred of each colour are needed. A square of grass (e.g. 15 m × 15 m) is marked out using string and pegs and the cocktail sticks are randomly distributed by the teacher within the square beforehand. A group of six students act as 'birds', feeding on the red and green 'insects'. They are allowed to feed for a 30-second period and to collect as many 'insects' as possible in this period. The results of their predation are scored individually. The process is repeated

Figure 8.3 The evolution of the horse

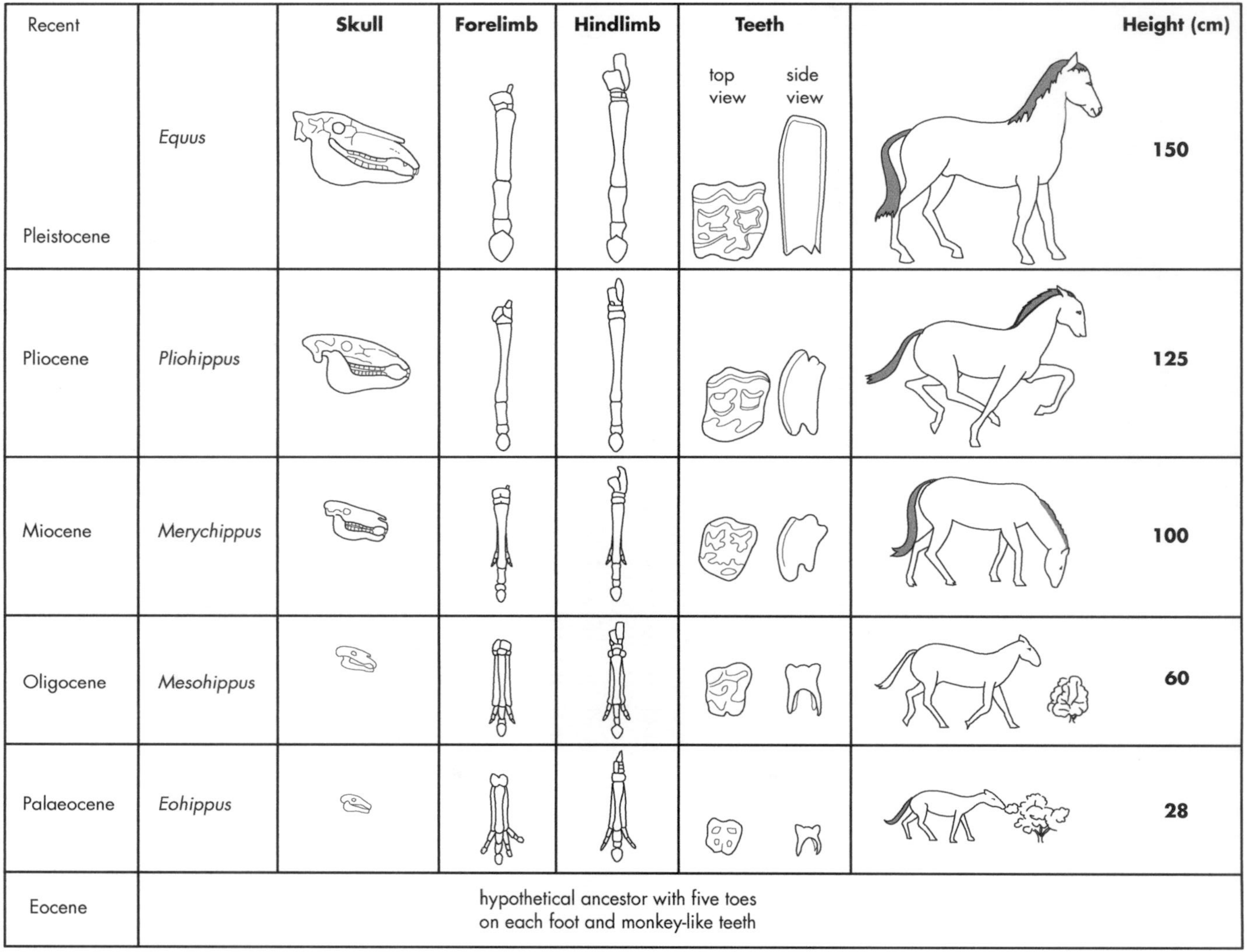

	Skull	Forelimb	Hindlimb	Teeth		Height (cm)	
				top view	side view		
Recent Pleistocene						150	*Equus*
Pliocene						125	*Pliohippus*
Miocene						100	*Merychippus*
Oligocene						60	*Mesohippus*
Palaeocene						28	*Eohippus*
Eocene	hypothetical ancestor with five toes on each foot and monkey-like teeth						

several times, placing the sticks in new locations, and the total number of prey items is totalled for the different 'birds'. Analysis of the results usually shows:

- the selective advantage that the green insects have in this environment
- that predators vary in their ability to detect prey (red/green colour blindness is particularly disadvantageous) and so on.

Students should be asked to think about the possible outcome if a similar situation occurred in the wild over a longer period time. The idea can be extended to think about different kinds of backgrounds and the effect on geographically isolated populations.

■ Modern examples

Useful and well-known examples are melanism in Lepidoptera (e.g. the moth *Biston betularia*) and predation by song thrushes on *Cepaea* (second-hand data about shells collected at anvils compared with frequency of different banding patterns in living populations of different habitats). Note that these are not instances of the formation of new species, but of selection leading to changes in the frequencies of different colour forms in populations of one species in each case. Such events are sometimes described as 'microevolution'.

Modern work on Darwin's finches is also useful to provide evidence of rapid selection (see below) as well as some additional historical interest.

■ Natural selection and evolution

One way to present the sequence of ideas is to use some chapter headings from the *Origin of Species*. Students could be provided with some selected quotations (see below) that can be discussed and interpreted from their earlier work on variation, genetics and selection.

The Origin of the Species by means of Natural Selection (1859)

- **Chapter 1 Variation under domestication:** 'Any variation which is not inherited is unimportant . . . but the number and diversity of inheritable deviations of structure . . . are endless.' (Darwin's examples were dogs, fowls, domestic pigeons.)
- **Chapter 2 Variation under nature:** 'The many slight differences which appear in the offspring from the same parents ... may be called individual differences. ... These individual differences are of the highest importance for us for they are often inherited ... and they thus afford materials for natural selection to act on ... .'

- **Chapter 3 Struggle for existence:** 'A struggle for existence inevitably follows from the high rate at which all organic beings tend to increase. ... Hence as more individuals are produced than can possibly survive, there must be a struggle for existence . ...'
- **Chapter 4 Natural selection; or the survival of the fittest:** 'This preservation of favourable individual differences and variations and the destruction of those that are injurious, I have called Natural Selection or the Survival of the Fittest. ... If variations useful to any organic being ever do occur ... individuals so characterised will have the best chance of being preserved and ... will tend to produce offspring similarly characterised.'

■ Controversial evolution

For some students, evolutionary ideas may pose problems because of religious views that they hold and hence this topic may be a controversial one. This is an additional reason for you to think very carefully about how you present the ideas, and it could be helpful to discuss your approach with colleagues in your department and have an agreed strategy. In some schools this may need to have the approval of the principal.

One idea that ought to be avoided is presenting Creationism or Intelligent Design as a valid scientific alternative to evolution. They are not, and many people believe that they should not even be discussed (let alone taught) within science lessons. For them, the correct forum for a discussion of this nature, if it is to be had at all, should be Religious Studies.

Other people may take a more conciliatory view. For example, it might be helpful to emphasise that it is quite possible for the theory of evolution and belief in a Creator to co-exist together. Science and religious views are not, of necessity, mutually exclusive.

If a school feels that this is an important issue to its students, then some form of collaboration between Science and Religious Studies might be developed to explore and debate the differences more fully. This would provide an opportunity to explain that creationism has no scientific basis, yet is a cultural 'worldview' (Reiss, 2008).

Further activities

- Visit the Natural History Museum website (www.nhm.ac.uk) and research the database on dinosaurs. This can be downloaded to an Excel file and used to classify the animals in various ways; to consider possible adaptations; to work out some possible food webs, etc.
- Research the possible reasons for the extinction of dinosaurs.

- Use an internet search engine and read about the discovery of modern coelacanths; discuss the significance of the coelacanth.
- Look at the reconstructions of some of the animals of the Burgess Shale in British Columbia (see Books; there is also a website: http://paleobiology.si.edu/burgess/). The University of California's website (www.ucmp.berkeley.edu/) includes details of other fossils. The trilobite section is interesting. All these provide vivid examples of animals that no longer occur on Earth and raise the question, 'Where did they go?'
- Read about Darwin's voyage on the *Beagle*. Darwin's works are all online at: http://darwin-online.org.uk/.
- The reception that the publication of the *Origin* received, the debate that followed and the reasons for the debate make this an obvious example to take to deal with the historical impact of a scientific idea. You might try some role play. Consider the present-day status of the ideas and how modern genetics provides additional information to support them. An account of the debate is at: http://users.ox.ac.uk/~jrlucas/legend.html.

Enhancement ideas

- Consolidation is required by giving students some further examples to discuss that are open to both Lamarckian and Darwinian interpretations. You might start with the legendary giraffe's neck or a weightlifter's biceps. Other possible ideas are to return to melanism (Did the industrial revolution and the increased soot production cause the melanic mutations?) or a host of other examples drawn from the literature or your own imagination.

Other resources

Books

Conway Morris, S. (1998). *The Crucible of Creation: The Burgess Shale and the Rise of Animals*. Oxford, Oxford University Press. Balances Gould's excitable view in the *Wonderful Life* (see below).

Conway Morris, S. (2003). *Life's Solution: Inevitable Humans in a Lonely Universe*. Cambridge, Cambridge University Press. Evolution explored by a scientist with a religious faith.

Darwin, C. (1859). *The Origin of Species*. A number of facsimile editions and reprints are available.

Dawkins, R. (1986). *The Blind Watchmaker*. Harlow, Longman. See Chapters 10, 'Classification' and 11, 'Lamarck, Darwin'.

Dawkins, R. (2005). *The Ancestor's Tale*. London, Phoenix. Readable Dawkins at his best.

Gould, S.J. (1980). *The Panda's Thumb*. Harmondsworth, Penguin. See pp. 65–71 for an account of Lamarck, Darwin and adaptation.

Gould, S.J. (1989). *Wonderful Life. The Burgess Shale and the Nature of History*. Harmondsworth, Penguin.

Maynard Smith, J. (1993). *The Theory of Evolution*. Cambridge, Cambridge University Press.

Poole, M. (1995). *Beliefs and Values in Science Education*. Buckingham, Open University Press. See Chapter 7, 'Darwin in context'.

Ridley, M. (1993). *Evolution*. Oxford, Blackwell Scientific Publications.

Ridley, M. (2003) *Nature via Nurture: Genes, Experience and What Makes Us Human*. London, Fourth Estate.

Ruse, M. (1982). *Darwinism Defended. A Guide to the Evolution Controversies*. Reading, Addison-Wesley. An account of Darwin's ideas and the Creationist positions.

Weiner, J. (1995). *The Beak of the Finch*. London, Vintage.

Articles

Brumby, M. (1979). Problems in learning the concept of natural selection. *Journal of Biological Education*, **13**, pp. 119–122.

Conway Morris, S. and Whittington, H.B. (1979). The animals of the Burgess Shale. *Scientific American*, **240** (Jan), pp. 122–133.

Deadman, J.A. and Kelly, P.J. (1978). What do secondary school boys understand about evolution and heredity before they are taught the topics? *Journal of Biological Education*, **12**, pp. 7–15.

Hey, J. (2009). Why Should We Care about Species? *Nature Education*, **2** (5), downloaded at: www.nature.com/scitable/topicpage/why-should-we-care-about-species-4277923. Date accessed 1/11/10.

Howarth, S. and Ingram N.R. (2010). Two cheers for biodiversity. *School Science Review*, **91** (336), pp. 23–25.

Lawton, G. (2005). Cannabis: Too much, too young? *New Scientist*, 26 March, pp. 45–46.

Reiss, M.J. (1988). Brown-eyed children may have blue-eyed parents. *School Science Review*, **69**, p. 742.

Reiss, M.J. (2008). Teaching evolution in a creationist environment: an approach based on worldviews, not misconceptions. *School Science Review*, **90** (331), pp. 49–56.

Theobald D.L. (2010). A formal test of the theory of universal common ancestry. *Nature*, **465**, pp. 219–222.

Internet

Arkive website www.arkive.org/ is an outstanding resource for video clips of endangered species.

bright hub www.brighthub.com/science/genetics/articles/41418.aspx looks at human variation.

Evolution MegaLab http://www.evolutionmegalab.org/ is excellent for banded snails.

Game of life www.bitstorm.org/gameoflife/ is an interesting game with an evolutionary twist. See http://en.wikipedia.org/wiki/Conway's_Game_of_Life for background information.

RSPB's bird identifier www.rspb.org.uk/wildlife/birdidentifier/.

A code of conduct is given in www.thewildflowersociety.com

All websites referred to in this chapter were accessed on 1/11/10.

Background reading

Berry, R.G. (1977). *Inheritance and Natural History*. London, Collins.

Briggs, D. and Walters, S.M. (1997). *Plant Variation and Evolution*, 4th edition. Cambridge, Cambridge University Press.

Driver, R., Squires, A., Rushworth, P. and Wood-Robinson, V. (1994). *Making Sense of Secondary Science: Support Materials for Teachers* and *Research into Children's Ideas*. London, Routledge.

Engel Clough, E. and Wood-Robinson, C. (1985). How secondary students interpret instances of biological adaptation, *Journal of Biological Education*, **19** (2), pp. 125–130.

Society of Biology (2009). *Biological Nomenclature,* Fourth Edition. London.

Lucas, A.M. (1971). The teaching of 'Adaptation'. *Journal of Biological Education*, **5**, pp. 86–90. An old reference, but still of value.

Osborne, R. and Freyburg, P. (1985). *Learning in Science: the Implications of Children's Science*. London, Heinemann.

9 Ecology

Susan Barker and David Slingsby

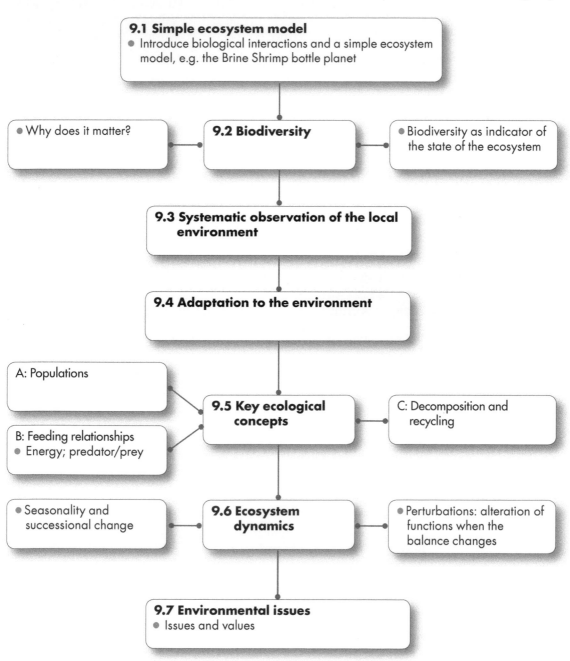

9.1 Simple ecosystem model
- Introduce biological interactions and a simple ecosystem model, e.g. the Brine Shrimp bottle planet

- Why does it matter?

9.2 Biodiversity

- Biodiversity as indicator of the state of the ecosystem

9.3 Systematic observation of the local environment

9.4 Adaptation to the environment

A: Populations

9.5 Key ecological concepts

C: Decomposition and recycling

B: Feeding relationships
- Energy; predator/prey

- Seasonality and successional change

9.6 Ecosystem dynamics

- Perturbations: alteration of functions when the balance changes

9.7 Environmental issues
- Issues and values

Choosing a route

Ecology is a branch of science that examines the interactions between living organisms and their environments. It presents teachers with a challenge, as not only is it viewed by some as the story of 'everything and everyone' but as students have already been exposed to ecological ideas in their earlier education, many believe they already know and understand the concepts. Start with finding out what they know first and then build on those understandings. The ecosystem is a fundamental unit of ecology and a good entry point is through studying interactions of organisms and their environment rather than to start by presenting the kind of overwhelmingly complicated diagram that graces many textbooks or through definitions and examples that lack context. The reason why generations of students have found the nitrogen cycle difficult is because each word on the diagram is the tip of a biological iceberg and there are too many difficult concepts per square centimetre on a single page to be absorbed in one lesson. Deepening the understanding of the whole is only as good as the deepening of the understanding of the components, such as biodiversity, populations, energy flow, recycling and ecosystem stability. Then the building up of a diagram of a complex ecosystem becomes a fascinating climax to the story.

Ecology is enlivened by an intrinsic fascination for biodiversity:

- Ecology assumes a knowledge of biodiversity.
- Biodiversity helps to explain why ecology is important.
- Ecology provides the understanding needed to conserve biodiversity.

The way for a student to begin to understand what the word 'biodiversity' means is not so much to learn a definition of it but to *see it* and *experience it* outside of the classroom. Such first-hand experience of reality can enhance the educational value of secondary sources such as video clips, web-based text, etc. It is important to highlight that knowledge of biodiversity can be used as an indicator of environmental quality, e.g. freshwater invertebrates are used for water-quality monitoring and lichens, air quality.

Field experience outdoors can be done even in the school grounds and is important in understanding interactions between one species and another, and between species and the environment. This too enhances knowledge gleaned from secondary sources and makes the all important connection between acting locally and thinking globally. Important secondary sources that can build on this experience include digital simulations that not only allow the study of ecological interactions to be extended but also allow scientifically derived hypotheses to be tested.

The teaching and learning of ecology need to take place in a spiral fashion. Small targeted field exercises strategically inserted into the spiral learning process in a way that provides progression can be particularly effective. The introduction of the ecosystem concept followed by an emphasis on animals and plants in a local environment and then the adaptations that enable them to survive provides a logical progression of concepts. This can lead to the exploration of ecosystem processes such as population dynamics, feeding relationships and recycling. It is important to show that ecological understanding of ecosystems is a scientific process that can be applied to explain what is happening, or to predict what may happen if a change occurs. The examination of some of these examples opens up possibilities for discussion of environmental issues. A really impressive message for students to take home from the study of ecosystems is that one apparently small change, either 'natural' or caused by human activity, can have far reaching consequences.

9.1 Simple ecosystem model

A teaching sequence

Many biology textbooks contain formidable diagrams of ecosystems. It is true that our goal is to introduce students to the fascinating complexity of ecological systems, but to make this accessible we need to keep it simple at the beginning. A deeper understanding develops by frequently revisiting the concept during consideration of its individual components.

Try to introduce and emphasise key concepts in a way that engages with your class. These concepts are:

- Organisms interact.
- Organisms are interdependent.
- Interactions and interdependence tend towards dynamic equilibria.
- Dynamic equilibria are not only easy to understand but are amazing!

The **brine shrimp model ecosystem** or 'Bottle Planet' (Figure 9.1) offers a very concrete route into ecology because each student can have their own ecosystem to look at, or you can have the class sitting around a single bottle planet. It is more visually effective if there is a powerful bench lamp shining on it. Where the students have little previous experience or prior conceptions, you can observe the brine shrimps and talk about their salt-lake habitat, their life cycle and their behaviour. Most students will be able to suggest why the brine shrimps are moving in pairs. You could mention that the shrimps are crustaceans and then discuss what other species they are related to and highlight that many people have had brine shrimps in similar bottles for years. You can tell your students that you never have to feed them or clean out the bottle, nor do you ever have to add extra brine shrimps to maintain the population even though every so often you see a dead one floating in the water. You then invite the class to discuss how this 'dynamic equilibrium' can exist. Alternatively, especially if each group of three to four students has its own bottle planet, you might like to provide a worksheet with questions that require observations to be made and then end with a class discussion. Finally, students could devise their own flow diagram to show how this ecosystem is a self-sustaining one.

Figure 9.1 The Bottle Planet Ecosystem. (Based on Dockery & Tomkins, 1999.)

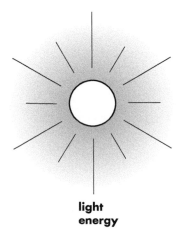

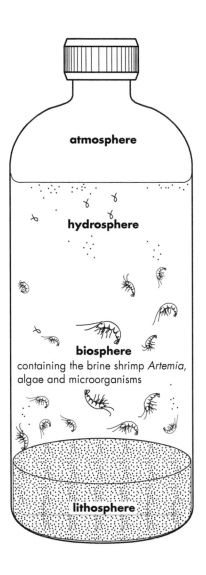

The story you are helping the class to arrive at is that brine shrimps feed on microscopic algae but the algae never run out because as fast as they are eaten, new ones are formed by multiplication and growth. In addition, oxygen never gets all used up because the algae recycle it through photosynthesis and when algae and brine shrimps die, there are bacteria in the system that decompose them and liberate the elements for reuse. All the system needs to sustain it is light energy. The bottle planet is described in the *Brine Shrimp Ecology* book by Dockery and Tomkins (1999) and it can be downloaded, free of charge, from the British Ecological Society website: www.britishecologicalsociety.org/educational/brine_ shrimp/index.php. As well as including all the instructions for setting up your own bottle planet, it also contains investigations in which the bottle planet can be used to study ecosystems at various levels of complexity. A Brine Shrimp Ecology kit, containing what you need to get started, including brine shrimp eggs and the all important inocula of algae and bacteria with which to 'seed' your first bottle planet can be purchased from Blades Biological: www. blades-bio.co.uk/. Once you have established your first bottle planet, you can seed as many others as you need (dried brine shrimp eggs can be bought in most pet food stores under the Latin name *Artemia* or the marketing name of 'Sea Monkeys').

The bottle ecosystem is a useful entry level activity as it helps you to determine the student's existing level of understanding and where to go next. You could discuss how in nature the shrimps are eaten by predators as fast as they reproduce but there are no predators in the bottle planet because it is very hard to get a flamingo in a lemonade bottle! Also, most students will have heard of photosynthesis and know something about mineral salts but they may find it hard to transfer their knowledge to the situation in the bottle planet. You may well discover a gap between knowledge and genuine understanding; for example, understanding the role of microbes in the processes of decay and applying this to recycling is often very shallow.

Before going into detail, take what the students already know and understand and weave this into an emphasis on interaction, interdependence and equilibrium.

9.2 Biodiversity

A teaching sequence

Biodiversity is one of those words that everyone has heard of, that everyone knows is somehow important but which most people do not quite understand the meaning of. But once someone has seen a video of life on a coral reef, freshwater invertebrates in a sample of clean stream water or the animal life in leaf litter from woodland, or made a visit to a marine aquarium they know what it is and what it is to be excited by it. Biodiversity is something to be *seen* rather than defined in words. Any attempt to define it as, for example, 'the variety of life both between species and within species', is somehow inadequate until you have experienced the awe and wonder of the real thing. It does, however, embrace a spiritual, cultural and ethical view of the world as well as a scientific and economic one, often being described as 'the library of life', 'drugstore' or even 'common heritage'.

In understanding biodiversity there is no substitute for at least some field experience. Some students may arrive in your class with experience of observation of biodiversity in primary school whilst others may not. Even if the students are already well prepared, their understanding of biodiversity needs to mature and progress throughout their secondary science experience, perhaps alongside other aspects of ecology. It is important to build on (rather than stifle) natural curiosity. Observation can be presented as an intellectually challenging aspect of secondary science teaching by:

- applying observation in unfamiliar situations
- using observation with predictions
- moving towards systematic and more sophisticated investigations
- integrating observation with other areas of the science curriculum
- linking observation to data analysis.

■ Making observations

A useful starting point is to pose questions to help students to make predictions. For example:

- How many *species* live in a tropical rainforest?
- How many *species* live in the school grounds (or other local habitat)?

The students could use secondary sources (library, internet, etc.) to find answers to the first question, then be invited to find out how many different organisms actually live in the school grounds. At its simplest this could be a class walk around the grounds, listing conspicuous species, progressing towards more specialised sampling to get a more complete picture, e.g. setting pitfall traps to find invertebrate species that are active only at night or hide when a class of 30 students violate the privacy of their habitat! Students are often surprised to find how long the list of species can be following some searching of a habitat:

- Bring samples of fresh soil into the classroom and examine them with a magnifier.
- Look closely at a patch of grass.
- Look at water from a pond or stream water under a microscope, or at the invertebrates it contains in a shallow white tray. For close examination with a hand lens, organisms may be captured in a wide-mouthed pipette.

Good hygiene is essential when handling soil and pond water. Avoid soil that is likely to be contaminated, e.g. from places where people walk their dogs. Also avoid ponds that are likely to be contaminated with chemicals, sewage or animal urine, especially if there are rats (carriers of Weil's disease) in the area. In any case a clean pond usually has the greatest and most interesting biodiversity and the risk it presents to students is no greater than that of any practical activity normally carried out in a science laboratory (and lower than many). Choose a site where students are very unlikely to fall into a pond or stream, and where the water is shallow enough so that even if they do they are unlikely to ingest water. Wash hands as soon as possible after working with pond or stream water or with soil. If a student experiences flu-like symptoms after working with pond water they should be advised to see their doctor just as a precaution. He or she will probably be found co-incidentally just to have flu but in the very unlikely event of them having contracted Weil's disease it can be readily treated if detected early.

Naming and classifying organisms is a natural progression from this activity and students are usually keen to do this. When a student realises that there are two or three types of beetle, they tend to want to know what to call them. Names give students mental pegs to hang things on and can lead to them realising that there are not two or three types of beetle in front of them but five or six! This indicates that you are succeeding in opening the students' eyes to the fascination offered by biodiversity. Concerns about identification need not and should not put you off taking students into the field, but it is important to consider how to overcome the problem.

■ Identification and the beginner: strengths and limitations

Some systems are much simpler to work with than others. An example of such a simple system involves only common soil invertebrates (such as spiders, beetles and centipedes) found using pitfall traps (Figure 9.2 on page 254) and identified using the easy-to-use field keys such as those of the Field Studies Council (www.field-studies-council.org/publications/) with pictures of soil invertebrates (Bebbington *et al.*, 1994) and 'Freshwater Name Trail' for invertebrates of streams and ponds (Orton and Bebbington, 1996). The keys, available for a range of habitats, are durable (being on laminated card) and inexpensive; thus a class set is not an unreasonable proposition. Students can easily use these themselves with a little help and encouragement.

You do not need to pretend to be an expert on invertebrate identification (it can be very liberating to discover that the students did not expect you to be anyway). You can work with the students, teach study skills and share your own fascination and curiosity. Perhaps the most important thing is to enjoy it yourself and let your enjoyment be infectious.

Wherever contact with living organisms or their habitats is made, it is worth mentioning at the outset that respect for life is paramount and that any specimens collected should be returned to their original source (alive and well!).

■ Identification and the expert: the advantages and the pitfalls

Inexperience in identification need not be a barrier to teaching good biology, but there is no disputing the fact that expertise can be useful. There are, however, pitfalls for the 'expert'. If the ecological story is mainly about trees, brambles, nettles, ivy, grass and bare ground it may be simply irrelevant and confusing to encourage (certainly to require) students to record obscure species as part of your ego trip. Also, it is not necessarily a good thing to tell the students everything when they can work things out for themselves. It is, as with all teaching, a question of sensitivity to where the students are at. When someone says, 'I think there are two sorts of these things with blue flowers – both look like speedwell', if you are there at the time and can say, 'Well spotted – the big hairy ones are germander speedwell and the one with smaller leaves is called slender speedwell' you will reinforce some fine discrimination skills. If you then add, 'and do you think they grow in the same sort of place or not?' then you are more than likely teaching good ecology.

■ Monitoring of water quality: an application of knowledge of biodiversity

Organisations responsible for monitoring water quality of rivers and streams employ people who are, effectively, professional pond dippers. Scooping up a sample of water or sediment from a river, stream, pond or lake and looking at the range and type of invertebrate species can indicate a great deal about the state of the ecosystem. Very high species diversity, including species with a high oxygen demand such as the larvae of the mayfly (species of Ephemeroptera) and of caddis fly (species of Trichoptera) indicate that the water is well oxygenated and of 'high water quality'. This is what you might find in clear mountain streams and the water is almost certainly fit to drink (although we don't recommend it!). If the water in a stream has low biodiversity, e.g. few species other than the sludge worm (*Tubifex tubifex*), there is almost certainly something wrong. Other species have intermediate levels of tolerance as shown, in a simplified form, in Table 9.1. When using this table start at the top. If there are no (or hardly any) larvae of mayfly, stonefly or caddis fly but there are greater water boatman and some of the water snail species mentioned in the table you would conclude that the water quality was fairly good but not so good as a pristine mountain stream. However, if there were some indicators of high water quality you would expect some of the other species indicating lower levels of water quality to be there as well and you would base your conclusion on the indicators of the highest level of water quality present. For further information on the use of biotic indices see more advanced sources such as www.ceh.ac.uk/products/software/RIVPACS.html.

Table 9.1 Indicator species of fresh water quality

Water quality	Indicator species
very high	larvae of mayflies, stoneflies, caddis flies and the freshwater shrimp
fairly good	greater water boatman, water snails (especially Jenkins spire shell, ramshorn, pea shell)
moderate/rather poor	alderfly, lesser water boatman, water cricket and leeches
poor	sludge worm, chironomid midge larvae, cranefly larvae.

If a river seems clean when someone goes to collect samples for analysis, yet has very low biodiversity, it could be because from time to time there is an intermittent discharge of pollution into the water. Chemical analysis would only show this if collected soon after the pollutant had been released. Looking at the biodiversity gives a longer term perspective and if there were problems would initiate a much more labour intensive (and more costly) series of chemical and microbiological analyses.

After an actual study of a freshwater habitat you could extend the scope of your visit by means of a computer simulation of a pond ecosystem, e.g. www.newbyte.com/uk/software/biology/.

9.3 Systematic observation of the local environment

A more systematic approach to observation is essential to allow predictions and hypotheses to be made. This moves us away from 'What is this?' to 'Why is there more of this here than over there?' and thus starts to explain why the distribution of organisms is not random – a key aim of ecology.

Try to encourage students to:

- be curious
- notice interesting patterns
- describe patterns more carefully
- formulate hypotheses
- make predictions
- test ideas.

Collection of data in a systematic fashion is often remembered for its repetitive nature rather than the purpose of the exercise; thus it is *vital* that a purpose is identified and made explicit. However, collection of good scientific data does have an element of repetitiveness in it. Students often find the precision very satisfying and quite like doing it as long as:

- it is clearly focused
- it is accurate – students can see that it is scientifically valid rather than crudely subjective
- a lot of data are collected quickly.

■ Good practice in fieldwork

All schools have policies that cover working outdoors with students. Yours will give you requirements or guidelines on staff/student ratios, first aid certification requirements, etc. Do not underestimate the importance of such information as you are legally responsible for the safety of students in your care. Being outdoors can be highly motivating, so as a teacher you need to think carefully about class management needs with excited students. Think about issues such as:

- What if a student needs toilet facilities?
- What if the weather turns inclement whilst you are in the field?

- What are the contingency plans in case of an accident, e.g. availability of a mobile phone – check that you would be able to get a signal. If a teacher has to take a student to hospital is he or she covered by insurance and will there be enough staff left to supervise the class?
- What if a student wanders off? Again, if a teacher has to go and look for them will there be enough supervision left?
- Have I advised students, in a letter that goes home to be signed by parents, of the arrangements, of how to be dressed (e.g. long trousers for working amongst brambles and nettles), to bring waterproof clothing, advice on appropriate footwear, and to bring their own insect repellent, sunscreen and hay fever medication?
- Have you checked whether any of the students have special needs or medical conditions?

It sounds a lot of hassle, but going through the processes once and establishing good protocols makes it much easier the second time. Careful planning makes it all go with a swing, particularly if it includes preparatory site visits not only by yourself but with your colleagues. Safety issues in fieldwork are often greatly exaggerated due to an extremely small number of well publicised incidents. Of course, safety is extremely important but statistics show that biology fieldwork is no more dangerous than any other form of science practical work (and less than much of it). Fieldwork is very motivating and educationally very worthwhile, so do not let the bureaucracy put you off.

Fieldwork also needs to be environmentally responsible, especially with regard to sensitivity to ethical and conservation issues and legislation on species protection. Make sure you gain permission from the landowner and keep to paths wherever possible. If you, as a teacher, demonstrate that you care for, appreciate and value the environment, then you are likely to instil these qualities in your students.

■ Pitfall traps

These are very easy to set up and can enable students to collect quantitative data as well as beetles! (see Figure 9.2) At one level, pitfall traps can be used simply to find out what is there. A grid of (say) 25 traps might be set by a class in a woodland and then another 25 in a different habitat, such as an open field. The class might have predicted that there would be more woodlice in the wood than in the field because there are more dead leaves for them to eat and the study should enable this prediction to be tested. For a more quantitative survey try the mark–recapture method to estimate population sizes in different habitats and/or at different times of year.

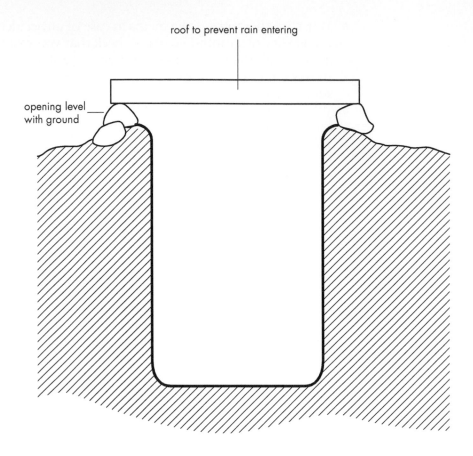

Figure 9.2 A correctly set up pitfall trap

roof to prevent rain entering

opening level with ground

■ Quadrats

Technically a quadrat is a sample area and is defined by a quadrat frame; in practice most people refer to quadrat frames as quadrats. You cannot possibly look at every square centimetre of a study area so a quadrat is a method of obtaining representative samples. Recording such representative samples is a means of *estimating* what the whole area is like. This offers a way of introducing the principles of statistical sampling to a science course. There are all sorts of different levels and approaches to the use of quadrats and thus subsequent data analysis can be tailor-made to meet the needs of your class's level and ability. If this is the first experience of sampling in the field it is important that *you* as a teacher keep it simple, select the appropriate method and present it to the students for them to apply in their own individual investigations or class exercise.

What shall I record?

You can record frequency, density or cover. At this level it is usually best if you introduce the exercise having already selected the sampling method and shown the students how to carry it out before inviting them to apply it. To be truly objective, quadrats should be placed randomly, using random co-ordinates. The picturesque idea of throwing a quadrat (frame) is not only dangerous but is never carried out as part of a serious scientific study.

Frequency

This is the probability of a species occurring in a randomly placed quadrat (see Figure 9.3). Essentially you record whether a species is present or absent in a large number of rather small quadrats, e.g. 10 cm × 10 cm, and each species is expressed in terms of the number of quadrats in which it was recorded as a percentage of the total number of quadrats used.

Density

This is the number of individuals per unit area and could be used, for example, to estimate the number of barnacles per square metre on rocks on a sea shore. It could also be used for estimating the number of species of plants in a pasture although it can be unsuitable where it is difficult to determine where one plant stops and another starts, e.g. grass (unless your students are obsessively meticulous). In such situations it is probably better to use percentage cover.

Figure 9.3
Recording cover using quadrats – subjective cover estimate

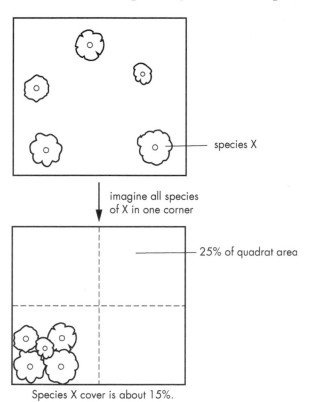

imagine all species of X in one corner

25% of quadrat area

species X

Species X cover is about 15%.

Percentage cover

This is the percentage of the quadrat covered by a species. There are two methods for its determination:

1 Cover estimation – here you make a subjective assessment (by eye) of the proportion of the quadrat covered by each species. The recorder can imagine that all the individuals have been moved into one corner of the quadrat before giving it a percentage cover value (Figure 9.3). Where the vegetation is layered with large plants such as trees and shrubs overhanging smaller plants, the total cover can be well over 100%. In woodlands one needs to look upwards and record tree cover before completing the recording of a quadrat.

2 Using a point quadrat – instead of a square frame lying on the ground, a point quadrat frame stands upright and has a row of evenly spaced holes through which a knitting needle can be inserted and allowed to fall (see Figure 9.4). Each species hit by the point of the needle as it falls is recorded in a tally. The needle may, for example, hit a blade of grass first and then a dandelion leaf at ground level. Both species would be recorded. The point quadrat frame can be placed randomly or regularly (say, every metre along a measuring tape). A score of ten hits on grass would thus be 100% grass cover, whereas dandelion, hit only four times, would score 40%. If the point hits the ground without touching a plant first then it is recorded as a hit on 'bare ground'.

Point quadrats are particularly suitable for investigations in short vegetation, such as the closely mown grassland of a school playing field. However, there are two disadvantages to point quadrats. Firstly, to get enough point quadrat data is more time-consuming than cover estimation. Secondly, point quadrats are unsuitable for use in tall vegetation.

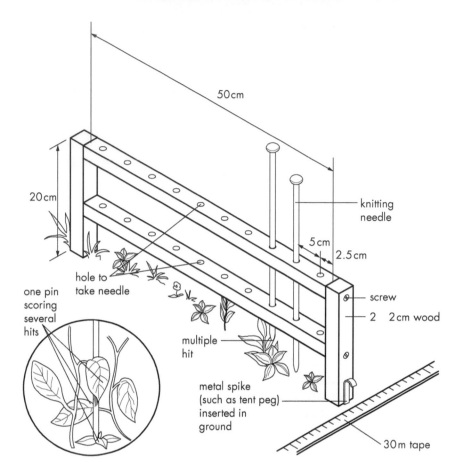

Figure 9.4 Using a point quadrat

A teaching sequence

■ When and where do I teach quadrat techniques?

The most effective way of teaching how to use quadrats is usually in the field when the students are motivated and keen to get going. A theoretical lesson on quadrat techniques is about as interesting as watching paint dry on a Friday afternoon, with equal learning outcomes!

Key points in using quadrats are:

- The focus of the investigation is the hypothesis rather than the methodology.
- Have a prediction that is to be tested – a clear focus.
- Use quadrats to collect relevant data.
- Collect sufficient data but not too much.
- Be consistent in sampling and recording techniques so that comparisons between studies can be made.

■ Making comparisons in grassland

Any grassland habitat would be suitable, e.g. a football pitch, lawn or meadows. At first sight students might consider this habitat to be all rather boring and 'just grass'. It will not take long for them to realise that it isn't all the same. There are may be numerous microhabitats at the site, e.g:

- football pitch: trampling in goal mouths
- cricket square: closely mown – weed-killer, extra fertiliser?
- shaded areas under trees
- close to path: low species diversity – recently disturbed?
- by river bank: high species diversity – moist soil and/or old established turf?
- long, straight, narrow strip with daisies – over buried pipeline?
- sloping area: better drained – drier?
- rough area: longer grass – mowed less frequently than the rest?

Working individually or as a group, suggest that the students select two places on the grassland that they feel are different, make a prediction and then test it by collecting data through the use of quadrats (they must, of course, be able to justify their predictions). Remember to show the students the quadrat methodology. In this example, a frequency procedure would be most appropriate.

10 cm × 10 cm quadrat placed randomly in each area 50 times. Each time just the species present are recorded as a tally.

$$\text{frequency for each species} = \frac{\text{number of quadrats with species}}{\text{total number of quadrats} \times 100}$$

If this is a preliminary exercise in which students are working independently, placing quadrats by laying two 30 m tapes at right angles and selecting points by use of random co-ordinates may be difficult. It may be easier for students to place their quadrats along a line at regular intervals. This is not strictly a valid procedure (since the quadrats are placed regularly rather than randomly) but at least the sampling is objective and will yield meaningful results. If you have a large class working individually you may not have enough tapes. One solution to this is to ask students to bring a long piece of string marked regularly at every metre along with an empty baked bean tin for pitfall trapping. Keeping the class wondering what the baked bean tin and the piece of string are for – until they get to the point where they need them – may increase their receptivity. It's all a matter of pedagogical psychology!

An example of a completed record sheet for this method is presented in Table 9.2. Encouraging students to emphasise the key features of the data by means of a summary table or chart will help them to acquire the important skill of seeing the wood for the trees.

Table 9.2 Recording sheet for using frequency to compare species composition of two contrasting habitats

	Shaded grassland			Unshaded grassland		
	Tally	Total records	Frequency (%)	Tally	Total records	Frequency (%)
Daisy	III	3	6	HHT HHT HHT	15	30
Grass	HHT HHT HHT HHT	20	40	HHT HHT HHT	15	30
Dandelion	HHT HHT	10	20	IIII	4	8
Rough hawkbit		0	0	HHT HHT III	13	26
Bare ground	HHT HHT	10	20	III	3	6

The hypothesis was that some species are more common in the shade and others in the open because some species of plants have adaptations that allow them to manage with less light than others.

■ Transect: changes and patterns from place to place

A transect is a line along which recordings are made in a systematic way in order to describe patterns and relationships. Across a woodland margin is just one of many places to record a transect as the habitat changes significantly over a very short distance. It will be very clear to the class that, as one passes into the wood, the vegetation and microhabitats change. There will be less grass but more brambles, possibly ivy and bare ground and, most obviously, the cover of trees overhanging the ground will increase. In the middle of the transect you will find typical woodland herbs but possibly also more common generalist plants such as nettles. As woodland communities are strongly influenced by seasons you may find relics of the previous season, e.g. if your fieldwork is in the summer you may find bluebell leaves and seed pods but not flowers. This is an excellent teaching opportunity to point out seasonality.

Concentrating on only a few species focuses the mind on the *changes* across the transect rather than *names* of large numbers of species.

■ Deciding what to do as a group activity: hypothesising and predicting

A class discussion is a good way to start thinking about making predictions and formulating hypotheses. Prompt the class to think about differences in light intensity and that this might be a critical factor in influencing the vegetation. This will possibly lead to a suggestion that it would be worth measuring the light intensity at ground level along the transect. Other things students might measure are: soil pH, soil and/or air temperature, humidity, etc. You as a teacher may know whether any of these factors will influence the change in vegetation across the transect, but they could and such speculation is all part of an investigation.

It is desirable to have a number of instruments to measure abiotic factors. Digital meters are available for all the variables you are likely to measure but older equipment is often more reliable and can be just as accurate. Purchasing useful equipment can add a great deal to the value of such an exercise. Data recorders/loggers with probes that collect data that can then be exported to computers for analysis with spreadsheets are a good way to measure light intensity, humidity and temperature (perhaps at different soil depths) over extended periods, e.g. a whole day or week. However, much can be achieved with simple equipment.

If you intend to return to a site the following day, the class could set up pitfall traps along the transect – perhaps there are more or different soil invertebrates in the humid and shady woodland than out in the open. A group of students thrust into the alien environment of a woodland soon warms to a transect and enjoys measuring things along it. You need not discourage such enthusiasm but it can be channelled along scientific lines within a hypothesis-generating and prediction-making culture. Your role is not to have all the answers but to have a few good ideas to get the students coming up with their own.

■ Getting on with the job

Once the students know:
- where to set out the transect
- how and what they intend to measure
- why they intend to measure it

they are ready to start. Remember to guide them as to how to measure vegetation cover, e.g. recording plants by percentage cover every 1 m using a 0.5 m × 0.5 m quadrat. Results can be recorded in a pre-prepared table (see Table 9.2). In 1 hour you could reasonably expect students to record percentage cover and abiotic data in about ten quadrats.

If a whole class is to work together on a single transect, make sure that everyone numbers the quadrats consistently. If a tape has been used to mark out the transect, start at zero metres and number the quadrats by the distance in metres from the origin. The quadrat 1 m from the origin is quadrat 1.

■ Collating data and analysing them by computer

There are things the students need to learn about how a team collects and collates data so that the data are internally consistent and reliable. It may help if you or one of the students co-ordinates the process so that it is completed as quickly and accurately as possible. 'Who did quadrat number 5?' elicits 'I did: bare ground 50%, grass 20%, nettles 10%, brambles 20%, light 550 lux, pH 6.5, humidity 80%, soil temperature 19 °C'. All the class can then enter the data on their own record sheets. If students have worked on their own, or in pairs, this collation will be much easier and they can proceed to analyse their own transect. A group of students, however, can cover a longer transect and this may mean that they identify more patterns.

Using spreadsheets on a computer to analyse their data adds to students' satisfaction and provides an excellent opportunity to introduce some ICT. The introduction of computers has had a significant impact on ecology, since large datasets can now be handled and subjected to statistical analysis. Much of this can be done now by using a spreadsheet such as Microsoft Excel, but using software often based on Excel but customised for field data can be tremendously advantageous. *Fieldworks* www.hallsannery.co.uk/software.html and Merlin www.heckgrammar.co.uk/index.php?p=10310 are both data-handling Microsoft Excel-based packages that can be downloaded free of charge from the internet. Both enable you and your students to plot kite diagrams (see Figure 9.5) and histograms, and carry out a range of statistical calculations relatively easily. *Fieldworks* also includes a database about common plant species and additional modules useful in teaching geography as well as biology.

■ Discussing transect data

Exploring what the data means can be done by gathering the group around a computer screen and pointing out, for example, how light intensity goes down as tree cover goes up. Perhaps soil temperature goes down and humidity goes up on entering the wood. And why does the pH change (if it does) along the transect? Why do some plant species grow at particular places along the transect?

Transects rarely prove anything; they demonstrate patterns and relationships that lead to new hypotheses. Often there are several alternative hypotheses invited by the relationship and these need evaluating. A written account of a transect can provide good opportunities for teaching and assessing data analysis and evaluation skills to various levels. Figure 9.5 presents sample data of a woodland transect using *Fieldworks*. Species are presented as kite diagrams and physical (abiotic) factors by means of bar charts.

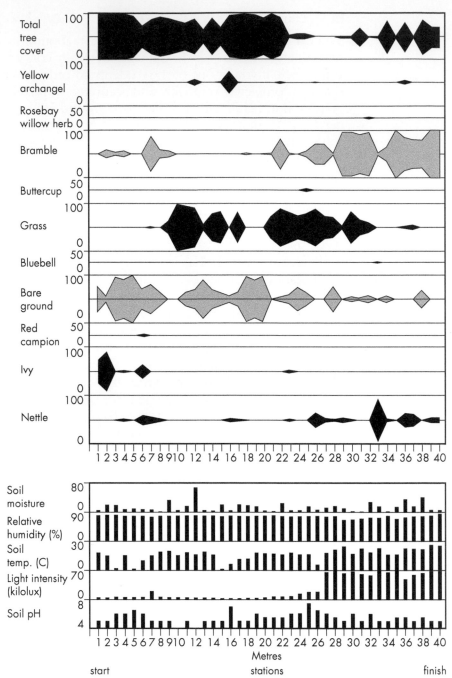

Figure 9.5
Woodland transect
data, presented
using *Fieldworks*

The key ecological trends shown in Figure 9.5 are as follows:

- There is a clear pattern of light intensity. Tree cover at the beginning of the transect casts a deep shade whilst at the other end of the transect, where there is less canopy, light intensity is higher.

- There is a pattern in soil pH and moisture. The variation in the data for soil pH suggests some experimental error – perhaps students did not insert the probe to the same depth?
- The data suggest that ivy could be shade-tolerant whilst bramble is more shade-tolerant than grass but less so than ivy.
- The temperature is cooler in the shaded part of the transect, but variations suggest recording errors. Was there a standard technique? For instance, was care taken always to insert the probe to the same depth every time a measurement was made?
- Humidity – this is highest in the shade, falling at the woodland margin, but rising at the end of the transect owing to the close proximity of a river. Humidity was measured using an electronic humidity meter but it can be measured in the field more cheaply using cobalt chloride paper carried in a boiling tube containing anhydrous calcium chloride (Figure 9.6). The time taken for the exposed paper to change from blue to pink can be an indicator of humidity and this is useful for comparative purposes.

Figure 9.6 A simple method for measurement of humidity

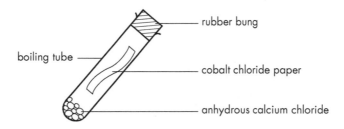

rubber bung

boiling tube

cobalt chloride paper

anhydrous calcium chloride

Further activities

- Individual investigations: students learn a lot from group work, but once they have had some experience they will be ready to work more independently. A transect or study involving a comparison of two sites or microhabitats is a reliable winner, but the best predictions come not so much from you as from your students!

9.4 Adaptation to the environment

Ecology is very much concerned with interactions between species and between species and the environment. Adaptation to the environment is a key biological principle that is important from an ecological point of view and is dealt with elsewhere (Chapter 8). However, the influence of the environment on a variety of plant and animal characteristics is often one of the easiest and simplest ecological investigations. Three examples of ecological studies are presented below. They are not particularly intended as exercises for you to carry out with your class but as illustrations of the kinds of activities you could get your students to do, either as class exercises or individual projects. They offer ideas that you may be able to adapt to different situations. No two habitats are quite the same.

A teaching sequence (case study)

■ Light intensity and leaf width in bluebells

Prediction: As the light intensity falls along a transect, the width of the bluebell leaves increases.

Scientific explanation: Wider leaves have a larger surface area and this is an advantage because this increases the ability to capture light energy.

Method: Select 100 bluebell leaves in the shade and 100 in the open at random. Use a ruler to measure the width of the leaves. Plot the results as overlapping histograms on a single sheet of graph paper by hand or use Excel, *Fieldworks* or Merlin.

Results: You may decide to introduce the normal distribution, depending on the mathematical interest of the students and on their previous experience. This, however, is not essential and better avoided if your students are not mathematically minded. There is likely to be some variation with overlap but it is highly likely that you will find a general difference between the two sites. The prediction, that the more light available the narrower the leaves, might well prove to be confirmed but it can prove to be the opposite way round and the baffled students might look to you for a clever explanation! A prediction doesn't have to be correct but it does have to be plausible and it may stimulate your class to come up with a better explanation. A key point might be that bluebells do most of their growth in spring before there are leaves on the deciduous trees. From a bluebell's point of view, what seems to be shady in May and June is actually well illuminated in March and early April, when it matters. In very shaded environments, e.g. under evergreens, photosynthetic capability is likely to be very much reduced all the year round; in these situations leaves may be considerably smaller overall – or bluebells may be absent altogether.

The various 'explanations' that come out of this study are hypotheses rather than firm conclusions. You may want to discuss how these hypotheses might be tested experimentally.

This exercise may also provide the opportunity to discuss ethical considerations. It can be carried out along a path without damaging the bluebell plants and without destructive sampling. It is a great opportunity to discuss with your class their own potentially destructive effects on the habitat. Working in a beautiful and interesting place elicits a combination of emotional, aesthetic and intellectual responses and makes the need to treat it with respect self-evident. This is far more effective than six 'boil-in-the-bag' reasons why we should conserve a rainforest (whatever that is), and it provides a basis for a real discussion about conservation of more exotic ecosystems.

Further activities

■ Case study: light intensity and leaf size in brambles

- On the basis of preliminary field observations similar to those in the previous example, a student predicted that the leaves on the shaded side of a bramble bush would have a larger surface area than those on the open side. The graph presented in Figure 9.7 confirms that bramble leaves in the shade have a larger surface area. Remember to point out that size of an organism or parts of it are only partially determined by environmental factors and that heredity is important too. Note that in this investigation, genetic diversity as a factor was eliminated by using shaded and open aspects of the *same* plant.

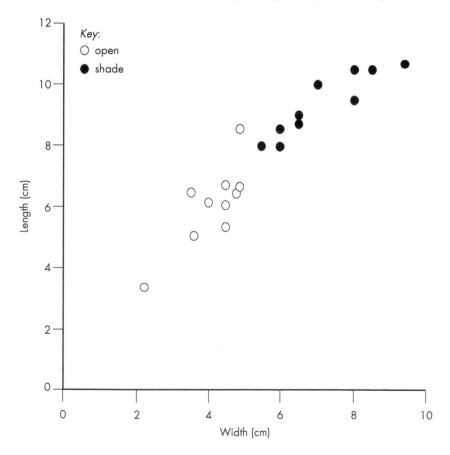

Figure 9.7 Size of blackberry leaves in open and shaded aspects, presented using the Warwick Spreadsheet System

- The graph on page 265 could be produced using Excel, *Fieldworks* or Merlin with a best fit line produced by regression. Or it could simply be plotted by hand with a best fit line drawn on it. Points to consider include:

- What does the relationship between length and width mean?
- Discuss the significance of the fact that the trend in leaf length relative to width is tending to level out at the leaf width above 6 cm.
- Would it be better to find the surface area of the leaves and plot this against light intensity? If so, how might we determine light intensity and leaf surface area?

■ Case study: soil moisture and soil organic matter along a transect at right angles to a stream through a hydrosere

- A hydrosere is the change in an aquatic plant community as open water is reduced by the growth of plants causing silting. As the soil builds up and the soil surface rises above the original water surface, the soil gets drier and different plant species move in. This successional series can be seen along a transect beginning at the open water and progressing into deeper and drier soil. Soil moisture can be measured by loss in mass on drying in an oven at 100 °C or by using a soil moisture meter and probe. Make sure that the probe is always inserted to the same soil depth before taking a reading. Soil organic matter can be measured by heating soil that has already been dried to a high temperature using a Bunsen or an oven until the organic matter burns – and literally goes up in smoke. The loss in mass of dry soil dried to constant mass (until all the organic matter has been destroyed and there is no further loss in mass) is equal to the organic matter that was originally present. It is usually expressed as a percentage of the mass of dry soil before heating to a high temperature.
- Soil moisture and organic matter were determined from ten soil samples collected at 1 m intervals along a line (transect) at right angles to a stream at Tarn Moss, Malham, North Yorkshire (Table 9.3). The relationship between soil moisture and soil organic matter at 1 m intervals along a transect at right angles to the stream is shown in Figure 9.8 and the relationship between organic matter and water content in Figure 9.9, both using the Merlin spreadsheet programme. The data were kindly provided by Robin Sutton of the Malham Tarn Field Study Centre (Field Studies Council).

- These data were collected as part of a transect study to describe the vegetation as you move away from the stream. You might expect the soil to get drier as you move away from the stream but, surprisingly, this is not the case here. The explanation is that the soil near the stream is low in organic matter and gravelly so that water drains out of it easily. This soil is very wet when it is raining and the level of the stream is high, but as soon as it stops raining it dries out quickly. Organic matter builds up because of vegetation, and organic matter retains water like a sponge. In dry weather the organic soil will stay moist much longer than the gravel near the stream and this provides the stable conditions needed for plant growth.
- Table 9.3 shows that the accumulation of organic matter is accompanied by a sharp reduction in pH. This because at Tarn Moss, the accumulation of organic matter takes the form of acid peat, despite the calcium-rich water from local limestone flowing through the stream.

Table 9.3 Soil moisture and organic content at metre intervals along a transect from a stream

Tarn Moss								
Sample	Distance along transect/m	Mass of crucible	Mass of crucible + wet soil	Mass of crucible + dry soil	Mass of crucible + burnt soil	% Moisture	% Organic content	pH
		W_1	W_2	W_3	W_4			
1 (by stream)	0	15.10	25.10	17.58	16.34	75.20	50.00	7.8
2	3	15.81	25.80	17.55	16.19	82.58	78.16	8.1
3	6	13.95	23.95	15.61	14.15	83.40	87.95	6.5
4	9	14.78	24.78	15.67	14.83	91.10	94.38	5.5
6	15	16.51	26.49	17.59	16.58	89.18	93.52	3.9
7	18	16.38	25.91	18.09	16.53	82.06	91.23	4.1
8	21	15.25	25.28	16.39	15.31	88.63	94.74	3.8
9	24	16.45	26.70	17.75	16.51	87.32	95.38	3.6
10	27	14.77	24.61	16.01	14.86	87.38	92.75	3.7

Figure 9.8 Soil moisture in relationship to soil organic matter content along a transect from a stream. What is the significance of the distribution of points between 90 and 95% organic matter?

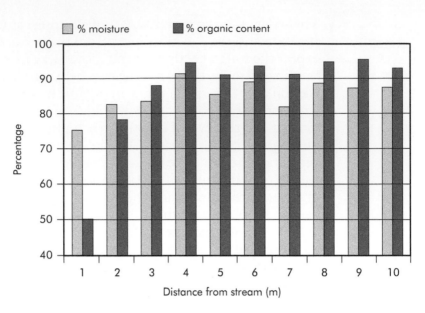

Figure 9.9 Relationship between soil moisture and soil organic matter content within 10 m of a stream

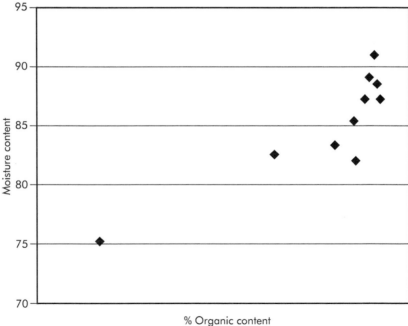

9.5 Key ecological concepts

The key ecological concepts which form this section are dealt with in most biology textbooks. The purpose of considering them here is to emphasise how they fit into real ecosystems, which you might visit with your students. What we are trying to do is to make them leap of out of the textbooks and apply them to explain what is going on in the field.

■ A: Populations

When a new bottle planet (Figure 9.1) is set up, often the water goes green as the algae are multiplying faster than they are eaten. This, of course, is because the algae multiply very rapidly. If we were to put more mineral salts in the water they might divide even faster and the water would become greener even more rapidly. If this did happen it would suggest that population growth was limited by mineral salt supply. If it did not then something else must be limiting – perhaps there is not enough light or, in the absence of brine shrimps and because the bacterial population hasn't got going, perhaps the algal population is taking up carbon dioxide faster than it is being replaced. Underlying all this is the principle that organisms reproduce and their populations increase until they are limited by availability of space, by the supply of a critical resource, by predation or by an outbreak of disease.

A teaching sequence

Keep a newly set up bottle planet (or maybe each student can have his or her own) in your classroom. Observe what happens for a short time every week and relate it to concepts and theories covered in class.

The brine shrimp population gets going more slowly than the algae. The eggs have to hatch, the shrimps have to reach reproductive maturity, mate, lay eggs, and the eggs have to hatch. Once the brine shrimp population has got going, the green colour fades as the shrimps eat the algae faster than the algae can reproduce. If you leave your brine shrimps, the increase in their population numbers may decline as they run short of food – or perhaps the algae aren't recycling oxygen fast enough. This may allow the algae to increase in numbers again – and the classic hare–fox oscillating population dynamics diagram found in so many textbooks happens in your laboratory. Limiting factors that might prevent the algae from increasing could be lack of fertiliser or supply of carbon dioxide. But once the bacteria in the system start to get going, feeding on brine shrimp droppings and dead brine shrimps, they recycle the mineral salts and the carbon dioxide. In this simple system all the factors rise and fall until – perhaps over several weeks – an equilibrium eventually becomes established. If we introduced a brine shrimp predator, such as flamingos (rather difficult in practice), it would take longer for the ecosystem to reach equilibrium. Achieving equilibrium in the natural world may take years, perhaps decades or centuries.

Finally, relate this to natural systems that the students have studied and to ones they have seen on videos such as rainforests. In the North Sea the equivalent of the brine shrimp is a small crustacean called *Calanus*, which is present in vast numbers, but not as vast as they used to be. They are fed on by small fish called sand eels and most of the bigger fish, such as herring and cod, feed on the sand eels. The *Calanus* feed (mainly) on single-celled planktonic algae, present in huge numbers. These algae start to multiply in spring as the sea warms up, and as they become more numerous the *Calanus* start to multiply. But some scientists think that, due to global warming, the *Calanus* are starting to hatch out earlier than usual, before the algae (which aren't as sensitive to temperature but start to multiply in response to day length) have got going, and so there aren't as many of them. This could be one of the reasons why there aren't as many fish caught and the price of cod is going up. Another reason might be that the *Calanus* are migrating north because of global warming. Of course there are other reasons such as over-fishing, but more of that later.

 Computer models www.newbyte.com/uk/software/biology/ originally designed for teaching natural selection at A level offer a lot of scope for students to explore adaptation, predation and populations in peppered moths, frogs and beetles. One of these programmes offers students the scope to add to their virtual frogs a gene that is toxic to their predators, or to introduce mimicry.

■ B: Feeding relationships

A teaching sequence

■ Food chains and food webs

A good starting point when teaching food chains and food webs is for students to describe feeding interactions in habitats from situations with which they are already familiar. In the school grounds a nibbled leaf, worm cast, even an empty crisp packet, all provide evidence of something feeding on something else. Animal behaviour also provides clues to feeding relationships, e.g. pecking for seeds, soaring/circling birds of prey and web-making to catch insects. Examination of the mouthparts of specimens caught in pitfall traps also offers evidence of feeding behaviours, e.g. chewing, sucking or piercing, and appendages for digging, grabbing, hooking or collecting. From this type of information, food chains and ultimately a food web can be constructed for the school grounds but it will only be a very small part of the actual complexity.

Getting students to construct food chains that explain the origin of a meal they have eaten can be a useful activity to get them to use secondary sources to link themselves with the global ecosystem. What, for example, do tuna fish eat and what are the conservation issues of commercial tuna fishing? What kind of food is fed to the animals that many people eat as food and what are the ethical issues that this knowledge raises? This links to nutrition (Chapter 2).

In Section 9.5A we began to piece together a food chain in the North Sea: Planktonic algae → *Calanus* → sand eels → herring and cod → humans. But sea birds such as puffins and kittiwakes feed on sand eels too. So a sudden collapse of such bird populations can indicate a shortage of sand eels and mean that a fishing crisis may be just around the corner. Young herrings eat *Calanus* rather than sand eels. Cod eat a variety of small fish. And seals and orcas (killer whales) also feed on the larger fish, competing with humans. Orcas also feed on seals (although this probably doesn't happen much in the North Sea). Here is the making of a good food web which your students can elaborate by using the internet.

■ Ecological pyramids and energy flow

One of the reasons why food chains are quantified and represented as pyramids (numbers, biomass and energy) is to demonstrate that the amount of energy transferred at each successive stage in the food chain decreases with increasing trophic level. Hints on teaching energy flow:

- Energy in biology is the same as energy in physics and chemistry.
- The Sun is the ultimate source of energy for most food chains.
- Energy is not recycled in an ecosystem – a continuous input is required to sustain it.
- Energy is 'lost' at each trophic level as heat to the environment.
- Food chains rarely have more than five links.

■ Constructing pyramids of numbers

It is possible for students to count producers, herbivores and carnivores in a sample from pitfall traps, from leaf litter or from a pond and thus construct a pyramid of numbers. This usually only demonstrates the principle of pyramid of numbers very crudely. In pitfall traps the carnivores may eat the herbivores before they can be counted! In leaf litter many of the herbivores, such as springtails, are very small and easily overlooked. In pond water the countless millions of microscopic algae on which most of the food chains rely for energy are not normally noticed at all. It is, however, well worth carrying out such an exercise because it makes the theoretical treatment, which must inevitably follow, much easier to understand by making some difficult concepts more concrete.

Good hygiene is essential when handling leaf litter and pond water.

■ C: Decomposition and recycling

Some students find decomposition a difficult concept to understand. Some reasons for this are:

- Microbes are too small to see.
- Lack of understanding that plants need carbon dioxide and mineral salts from the soil and that continued plant growth depends on the recycling of these substances.
- Substances such as carbon dioxide and mineral salts needed by plants rarely run out because they are continually recycled owing to microbial action with the help of detritivores such as earthworms and woodlice. Animals depend on plants to recycle oxygen and to produce food by photosynthesis.

It is thus important that students have already been introduced to plant nutrition (Chapter 2) and to microbiology (Chapter 10).

A teaching sequence

■ Decomposer investigations

Cellulase enzyme activity due to soil microbes can be demonstrated and measured simply by placing samples of moist soil in Petri dishes in incubators at 25–30 °C and recording how many days it takes small pieces of thin unglazed paper to disappear. (Most will go in a week.) Burying dead leaves in plastic mesh bags may be useful in specific projects but observing the subsequent decomposition usually takes too long for class practical exercises.

Decomposition and recycling

Detritivores are animals, such as earthworms and woodlice, that play a part in decomposition. When dead plant material passes through a detritivore it is partly digested and absorbed but what remains comes out in the droppings (woodlice fed only on carrots produce orange faeces). Digestion of the remains is then completed by soil microbes. Students often find this interesting and most investigations can be completed in 3 days.

The investigations can be set up to examine food preferences in woodlice and earthworms. Essentially, you collect leaf litter and identify the species (works well with oak, sycamore, ash, holly, beech). Cut 2 cm squares of each type of litter and place in an alternating sequence in a Petri dish (line with moist filter paper); mark the position of the fragments (they sometimes get moved around by the woodlice but for the most part remain in their original location); place two woodlice in the Petri dish; put in a cool dark cupboard and leave for 2–3 days. Remove the woodlice and return them to their natural home. Stick the fragments of leaf litter on to graph paper. By counting squares on the graph paper you can work out how much of each type of litter has been eaten and so work out woodlice food preferences. In most cases the woodlice will select the most palatable fragments first, usually partly-rotted leaves with a high nitrogen, moisture and sugar content, and will only consider drier, tougher leaves later. If you leave the experiment too long the woodlice will eat everything, so timing is quite important.

Most nutrient cycles contain some very difficult concepts; be careful not to teach more theory than necessary. You do not have to teach every nutrient cycle for students to get the idea of cycling. The overriding challenge is to find exciting and motivating ways to teach nutrient cycling.

9.6 Ecosystem dynamics

Ecology is about studying animals, plants and microbes interacting with their physical environment as systems – ecosystems. These systems sometimes remain remarkably similar for a surprisingly long time whilst others undergo changes before your very eyes. These changes (and sometimes the lack of them) constitute ecosystem dynamics.

A teaching sequence

■ Seasonality

The fact that the biological world around us changes with the seasons has long fascinated not only natural historians but also artists and poets. The traditional primary school 'nature table' has still much to commend it, as have wall posters about nature and the seasons. The bluebell examples on page 264 illustrate how the seasonal component can be brought into more systematic investigations. It is possible to make observations in school grounds concerning the seasons although the time to do this is usually limited within the constraints of the school timetable (more possibilities with project work). Videos and digital photographs in PowerPoint presentations of familiar habitats at different times of the year can be very useful.

■ Succession

Succession is an ecological process occurring before our very eyes and, since it is more concrete than concepts such as nutrient recycling, energy flow and ecosystems, it offers a key progressional link between reality and abstraction. Succession is a concept not often emphasised in the secondary curriculum but it is worth incorporating into a teaching scheme because it helps to ensure that a lot of the rest of ecology makes sense. A good way to study (secondary) succession is to clear a plot in the school grounds so that there is a patch of bare soil; records about this can be kept over months and years, and a database including quadrat data and photographs built up by successive classes. If sufficient space is available, a new plot may be cleared each year and successional progress compared with older plots. The pace of change will be rapid to start with but will probably slow down as the ecosystem tends towards stability. This notion of successional processes tending towards stability is an important one providing that one avoids being simplistic. The stability is relative, ecosystems rarely stay the same for long but on reaching relative stability, on reaching a 'climax community', change does not stop – it just tends to be very slow in comparison to what happens in the first year after the land was cleared.

Students are likely to be interested to learn that Britain was once largely forested because of its climate and that absence of forest today is due to human activity over the past 4000 to 5000 years. There are still plenty of trees even in an urban landscape, and what we think of as garden birds, such as blue tits, are actually woodland birds. Waste land rapidly becomes overgrown with brambles and trees such as elderberry. Old railway sidings are often taken over by silver birch – one of the earlier tree species in a succession back to woodland. If, due to some disaster movie scenario, Britain were to become largely devoid of human population much of our cities would, like lost Inca cities of South America, be swallowed up in forest – by successional processes.

The forest that regenerates might not be the same as it was 5000 years ago, since environmental conditions have changed and many new species of plants have been introduced from overseas. The impact of humans on the landscape, flora and fauna of Britain has been huge and this puts our current concern about biodiversity in perspective. Nor is climate change new – 12 000 years ago much of Britain was covered in ice with tundra in the south. As the global warming that ended the last glaciations (Ice Age) melted the ice, first pine and then oak forest spread back across dry land, which is now below the English Channel. Changes in temperature and rainfall, and then the impact of humans, meant the vegetation was always changing yet always tending towards a (changing) equilibrium.

The progression from bare soil to forest would take more than the lifetime of a student but students can observe parts of the process within a year. They can extend their experience by looking at last year's quadrat data (long-term monitoring is an important aspect of ecological research) and they can relate what they see happening to what you tell them about the theory. In many places succession leads to regeneration of woodland and rainforest destruction does not always lead to desertification.

The humble plot that you may have cleared in your school grounds in order to witness the first stirrings of successional processes can be the focus for a lot of good thinking. Ecosystems can be destroyed by humans, yet they can also recover by natural ecological processes. Many human activities threaten ecosystems and may upset balances but we can predict the outcomes if we understand how ecosystems work. Ecosystems can be more robust than they seem.

■ Perturbations

The notion that natural ecosystems tend towards stability can be very useful at secondary school level even if it can sometimes be an oversimplification. A perturbation is something that upsets this relatively stable system. Perturbations can happen naturally but human activity can often be the cause of them. This section examines the way in which the science of ecology helps not only to explain ecological events but also to predict the consequences of environmental change.

■ Food chains and food webs – bioaccumulation of persistent substances

Knowledge of food webs helps to explain why persistent substances introduced to the environment at low levels turn up at a high, and sometimes lethal, level in top consumers. Examples of such persistent substances include: the pesticide DDT; radionuclides, e.g. caesium 137; and heavy metals, e.g. mercury.

■ Ecological knock-on effects in a food web

A food web diagram can be used to predict:
- the effect of the removal of a single species from a web, e.g. through hunting or overfishing
- the effect of the introduction of an alien species, e.g. mink
- the impact of management regimes, e.g. grazing, burning.

On page 271 we observed that a collapse in the sea bird population of species such as puffins can be a warning that the whole North Sea ecosystem is under stress due to collapse of the sand eel population. The sand eels were being overfished to provide feed for domestic animals such as pigs. This threatened not only picturesque species such as puffins but the commercially important fishing of cod, herring and halibut. The annual value of cod caught in the North Sea is over 14 million pounds and provides many people with employment.

On page 251 we considered how the disappearance of mayfly larvae from a stream may mean that this is an ecosystem in trouble. A clean stream with high biodiversity can absorb a certain level of human sewage, its microbes can break this down and recycle the valuable nutrients. However, once the pollution gets beyond a tipping point the system becomes unstable, unable to cope with the added organic matter and other pollutants, its biodiversity drops, it can no longer support fish and possibly becomes a health risk.

■ Nutrient cycles

Climate change – a perturbation of the carbon cycle?

Climate change and global warming are contexts that make the carbon cycle interesting because they illustrate how a stable system can be influenced by humans with potentially enormous consequences – it's real life stuff! There are facts, hypotheses, predictions and falsities involved; thus providing interesting contexts to educate students in critical analysis and consideration of a variety of scientific approaches. One of the points of teaching climate change has to be to encourage your students to appreciate that what they hear about or read in the media or are taught is not all certainty: they need to be 'information literate', i.e. to be able to evaluate information, to distinguish between facts and speculation.

Some suggestions for teaching climate change are:

- The greenhouse effect is a natural phenomenon essential for life on Earth, without it the Earth would be too cold to support life.
- The carbon cycle shows how burning fossil fuels can lead to an increase in atmospheric carbon dioxide concentration. This has led to predictions of an enhanced greenhouse effect likely to cause global warming. There is evidence of an increase in atmospheric carbon dioxide and of increases in global temperatures but the link between them is not considered proven by all.
- There is clear evidence that the concentration of carbon dioxide in the atmosphere has increased since the 1950s, and probably since the industrial revolution in the eighteenth century, and the burning of fossil fuels during the past 200 years has almost certainly contributed to this. It is also indisputable that we are in a period of global warming – although this has happened before, e.g. at the end of the last Ice Age, long before humans

started burning fossil fuel. There is clear scientific consensus about human-induced climate change but there are some areas of contention.

- Is burning fossil fuel the *main* cause of the current global warming or is this largely caused by natural phenomena over which humans have little control?
- If burning fossil fuel is a major factor, could we reduce CO_2 emissions enough to make any difference – or is it already too late?
- If a drastic reduction in CO_2 emissions could make a difference, would it be politically realistic to expect this to happen? Should we devote our efforts to coping with climate change?
- There is plenty of evidence that burning fossil fuel *might* contribute towards global warming and the consequences of global warming *are likely to have a significant impact on human populations, the severity depending on where in the world you live.* There are also other advantages in reducing our dependence on fossil fuel (e.g. political considerations and the fact that fossil fuels are going to run out). Therefore these other reasons, combined with the serious possibility that reducing CO_2 emissions *might* reduce global warming, add up to a good argument for reducing the use of fossil fuel.
- When tropical rainforests are chopped down it is true that in the short term there is a sudden pulse of carbon dioxide released into the air from the processes of decay and burning. However, in the longer term less photosynthesis means there is less organic matter to be broken down into carbon dioxide. Thus in the long term a rainforest is a balanced ecosystem and it is not a net absorber of carbon dioxide. Basically, what goes up must come down.
- Ozone depletion and global warming are both global issues but the two are almost entirely separate and should be treated as such. However, it is useful to point out that because of action taken by governments based on scientific evidence, ozone depletion is no longer a major issue of concern.

Eutrophication – a perturbation of the nitrogen cycle?

In a normal freshwater habitat the reason why accumulation of algae does not take place is because algal growth is limited by inorganic nutrient supply (either nitrate or phosphate) and by grazing – small animals in the pond feed on the algae as fast as the algae reproduce. In the nitrogen cycle, biomass, such as that of algae, can be in equilibrium – the rate of production of new biomass is limited by the rate of decay of dead organisms (releasing available nitrogen). When a large amount of nitrate enters a lake or river from fertiliser run-off or from sewage treatment plants, the algae, whose growth has until now been limited by nitrate supply, increase in numbers very rapidly, much faster than animals can eat them. This leads to an accumulation of biomass that, after death, is decayed by bacteria. The normally low numbers of bacteria in an unpolluted stream now increase rapidly and use up much more oxygen than usual, which can cause the death of fish and other animals, such as stonefly nymphs, that need a lot of oxygen.

A final point to note about equilibria in ecosystems is that these are the result of a balance between many different factors all pushing in different directions. If we reduce the pressure from one direction, generally the *whole* equilibrium shifts.

■ Virtual ecosystems through computer modelling

In much of this chapter we have emphasised the importance of real outdoor experience when teaching ecology at this level – indeed, at any level. The reality of life in school means that much of the learning will nevertheless be indoors, but this can be massively enriched if there is at least some field experience and this experience is woven into the course as an integral and essential aspect of it. Using a bottle planet (Figure 9.1) becomes an integral part of teaching when it helps to make sense of the real ecosystems your students work in and these, in turn, help to make sense of more exotic ecosystems studied from secondary sources. Computer programs are no substitute for real field experience but they do become powerful teaching aids when they extend hands-on experience.

You may only have a day or two on a real rocky shore but Virtual Tours on the BES website www.britishecologicalsociety.org/educational/fieldwork/virtual_tours.php (also available on www.takeyouvirtuallyeverywhere.com/) will allow you to back up your field expedition with a virtual visit to a range of rocky shores in different parts of Britain. If your field work was on sand dunes or heather moors, virtual tours will enable you to extend your visits and also compare your site with others in Britain. After spending time in the British countryside you might like to take the odd virtual tour in a desert in the south west of North America. The virtual tours even contain high-quality photographs of quadrats along a transect, which are so good that you can identify and record the plant species just like in a real one.

A computer simulation called 'Rocky Shore Ecology' supplied by Newbyte software www.newbyte.com/uk/software/biology/ offers you a chance to compare the rocky shore you studied in Britain with one Down Under. An Australian rocky shore has different species, yet the way it works ecologically is similar to ones in the northern hemisphere. Newbyte also supplies computer models of Australian woodland, which students can manipulate in a way that they could not do in a real woodland back in Britain. The principles are the same, but the students find themselves asking what occupies the ecological niche of a wombat back home in Britain. Similarly, Newbyte supply a computer model of a pond, which enables students to manipulate the pond environment and test ecological hypotheses.

There are serious points to be made here. Modern ecology uses vast and complex computer simulations to make predictions and test hypotheses about whole ecosystems. Many of the predictions are about the ecological effects of climate change and allow one to predict the effect of an average increase of, say 1 °C or 5 °C, on a tropical rainforest. Such models are not perfect, but as more and more data are obtained and the models are improved they are tending to be more and more reliable.

9.7 Environmental issues

Ecology provides a great opportunity for biology teachers to raise awareness of global environmental issues. The relationship between teaching about environmental issues and ecology as science can sometimes appear confused and it's a good idea to distinguish between the two. This can help deepen an understanding of the nature of science and scientific evidence.

Ecology:

- is a branch of science with its roots in biology
- is objective, experimental and predictive
- is interdisciplinary with links to biology, chemistry, physics, earth sciences and geography
- includes concepts usually specified in science curricula.

Environmental education, Education for Sustainable Development or education for sustainability and Global Citizenship:

- helps people to understand, to appreciate, to care for and to enjoy the environment
- includes aesthetics, issues and values as well as scientific knowledge
- is often cross-curricular: it brings together elements of science with social science, policy, social justice, geography, art, aesthetics, language, mathematics, ethics, religious knowledge, sport, outdoor pursuits, etc.
- is often not part of the statutory curricula.

A teaching sequence

■ Teaching environmental issues: developing critical thinkers

Being a critical thinker is an important life skill for a scientifically literate person and examining environmental issues provides us with a great opportunity to develop these skills. The ability to think critically is essential if individuals are to live, work and function effectively in our current and changing society. As a teacher we can provide contexts for which there is no clear-cut right or wrong answer and encourage students to judge what would be sensible or reasonable to do in a given situation. Such tasks will only develop critical thinking if they invite students to assess the reasonableness of plausible alternatives. It is really important that students make their own decisions and reach their own conclusions as they find this highly motivating; also, if we as teachers tell them *what* to do, this leaves us open to criticisms of indoctrination. In addition, environmental issues can enhance the integration of scientific knowledge and connect students to real world issues. Teaching using a critical thinking approach is quite different to enquiry approaches to science teaching where the *quality* of decision-making is less important than judgements about scientific evidence.

In Section 9.6 we showed how we can apply knowledge of ecology to understand what happens when ecosystems become disturbed, all too often as a result of human interference. Ecological knowledge can also be used to work with nature. A well-managed ecosystem can provide society with resources year after year in a sustainable fashion, for example:

- commercial sea-fishing using knowledge of population dynamics to calculate the maximum sustainable catch from an area of sea
- use of biological control as an alternative to chemical pesticides, e.g. parasitic wasps to control whitefly in greenhouses
- growing wood as a renewable resource of biomass to use as an alternative to fossil fuels, e.g. fast-growing willows are used to provide fuel for generating electricity
- managing forests for sustainable timber supply
- managing reserves for wildlife and landscape quality for tourism
- conserving habitats and biodiversity for scientific and medical reasons
- organic farming for sustainable agriculture.

Jared Diamond provides a very useful account of how human practices can push ecosystems beyond the point of no return in his book *Collapse*. He identifies five factors that contribute to collapse of human societies: climate change, hostile neighbours, collapse of essential trading partners, environmental problems, and failure to adapt to environmental issues.

He also lists 12 environmental problems facing society that need ecology to help us understand the complexity of the challenges and provide scientific evidence to move forward with solutions:

1 Deforestation and habitat destruction
2 Soil problems (erosion, salinisation and soil fertility losses)
3 Water management problems
4 Overhunting
5 Overfishing
6 Effects of introduced species on native species
7 Overpopulation
8 Anthropogenic climate change
9 Increased per-capita impact of people
10 Build-up of toxins in the environment
11 Energy shortages
12 Full human utilisation of the Earth's photosynthetic capacity.

Using these problems as contexts for teaching environmental issues, whilst helpful, also needs focus. A useful pedagogical model to use is based on Nikitina (2006):

1 Contextualisation – tell a 'story', use a newspaper article, a website, YouTube clip, etc. This grabs students' interest and helps root the issue in many disciplines.
2 Conceptualisation – make explicit the links to ecological concepts in the curriculum.
3 Problem-centring – use specific issues in the local environment that require students to think critically and offer potential solutions.

A final point for emphasis about teaching ecology: this subject is about living things in their natural environment. Sensitive use of living organisms in the field and laboratory in your teaching can increase the sense of awe and wonder at biodiversity and also cultivate respect for living organisms. Wherever possible, within constraints of health and safety and animal welfare issues, do try it!

Other resources

Background reading

Books that will really help you with teaching ecology are:

Chapman, J.L. and Reiss, M.J. (2012). *Ecology: Principles and Applications* (3rd edition). Cambridge, Cambridge University Press.

Cotgreave, P. and Forseth, I. (2002) *Introductory Ecology*. Oxford: Blackwell Science Ltd.

Some useful internet support for you and for your students

iSpot: www.ispot.org.uk/ is produced by the Open University, and if you (and your students) want to deepen your knowledge of biodiversity this could be the website for you. It includes a blog where subscribers share observations and insights through a news section and a forum, items on groups (e.g. amphibians), keys for identification and surveys you can take part in. iSpot helps you to get into knowing more about plants and animals.

ARKive: www.arkive.org/ is an exciting website providing up-to-date information about plant and animal species, including many of those threatened with extinction. It has high-quality photographs and video clips of wildlife.

The British Ecological Society website has an education section www.britishecologicalsociety.org/educational/index.php with useful information on field work, and resources such as a wall charts on ponds are packed with information about how ponds can be developed in school grounds, specifically for teaching about ecology. At the time of writing, the Society and the National Science Learning Centre (York) are developing a series of short modules on the biological background (mainly for science teachers) to common organisms encountered in fieldwork such as earthworms, freshwater invertebrates, native trees, common playing field plants, snails and slugs, and woodlice. It is worth keeping an eye on this website for further developments.

The Woodland Trust and Nature Detectives have good material to download as well as projects for older students.

The Field Studies Council (FSC): www.field-studies-council.org/ publish laminated identification charts, and more advanced aids to identification. They also provide training courses on teaching in the field, including safety management. If you are fortunate enough to be able to, you could take your class for a residential field course at one of FSC's field study centres or at one of the many independent field study centres around Britain (see the website of the National Association of Field Study Officers (NAFSO) www.nafso.org.uk/.

TIEE: Teaching Issues and Experiments in Ecology: http://tiee.ecoed.net/ is a peer-reviewed publication of ecological educational materials by the Ecological Society of America.

References

Bebbington, A., Bebbington, B. and Tilling, S. (1994). *The Mini-Beast Name Trail: A Key to Invertebrates in Soil and Leaf Litter*. Preston Montford, Field Studies Council.

Diamond, J. (2005). *Collapse: How Societies Choose to Fail or Succeed*. New York: Viking Press.

 Dockery, M. and Tomkins, S. (1999). *Brine Shrimp Ecology*. London: British Ecological Society. Available from www.britishecologicalsociety.org

Fieldworks (1997) produced by Hallsannery Field Centre, Bideford, Devon EX39 5HE. (Runs on PC.)

Nikitina, S. (2006). Three strategies for interdisciplinary teaching: contextualizing, conceptualizing, and problem-centring. *Journal of Curriculum Studies,* **38** (3), pp. 251–271.

Orton, R. and Bebbington, A. (1996). *The Freshwater Name Trail: A Key to the Invertebrates of Ponds and Streams*. Preston Montford, Field Studies Council.

10 Microbiology and biotechnology

Roger Lock

10.1 What are microbes?
- Brainstorm
- Newspaper cuttings
- (Algae, fungi, bacteria, protozoa, viruses)

10.2 What do microbes look like?
- Microbe gardens
- Microscopic examination
- Size concepts

10.3 Where do you find microbes?
- Growing microbes
- Differences between fungi and bacteria

Why are microbes important?
- (Useful microbes, harmful microbes)

10.5 What do microbes need to grow?
- Broth experiment
- Food preservation

10.4 How do you handle microbes safely?
- Taping and labelling plates
- Incubating plates
- Pouring and inoculating plates

10.6 How are microbes involved in a healthy life?
- Hygiene and food preparation
- Clean water
- Digestion
- Microbes and disease
- Antibiotics

How are microbes involved in genetic engineering?

10.7 How are microbes involved in biotechnology industries?

Choosing a route

Finding out what microbes are, what they look like and where they are found is basic work accessible to 11 to 12 year olds of all abilities. Understanding these basic ideas is also a prerequisite to the content of other boxes in the flowchart and should determine students' early experiences of microbes and biotechnology.

Practical work is an important element of students' early experiences of microbes, and safe handling of microbes by teachers and students alike is essential for work that involves using cultures of fungi or bacteria. Work demanding a significant level of manual dexterity is perhaps best tackled with older students, say 13 to 16 year olds, but that outlined in boxes 10.1 to 10.3 can provide a basic framework for study with11 to 13 year olds.

The principles involved in personal and domestic hygiene arise from an understanding of the control of microbe growth, e.g. keeping food in the fridge slows down their growth but does not kill them. In addition, when exploring applied aspects of microbiology it is most effective to start with applications closest to students' experiences, as they will more readily relate to them, before going on to study the biotechnology in industry and genetic engineering.

The importance of microbes pervades work at all levels and has a place in many of the other boxes. For this reason it is best addressed at a range of levels, in different sub-topics, as and when appropriate, but it also provides a convenient heading under which work from all other boxes can be drawn together and summarised.

There are many reasons why microbiology should be given a prominent place in the biology curriculum. Microbes:

- are socially, economically and medically important
- are central to research in genetic engineering and developments in biotechnology industries
- play a key role in personal, public and domestic hygiene
- are important in food production
- cause diseases and help control them
- are essential in cyclical changes (e.g. carbon cycle, nitrogen cycle)
- help herbivores to digest their food.

A final reason for studying microbes is that they can be used to demonstrate a wide range of biological principles and processes, such as the S-shaped growth curve, photosynthesis and respiration, as well as illustrating the properties of living organisms; indeed, as some students think that microbes are not living things, using microbes to show that they move, excrete, respire, reproduce, react to stimuli and grow can be an effective way of creating cognitive conflict.

10.1 What are microbes?

Previous knowledge and experience

Microbes and biotechnology are in the media a great deal but the exposure they get is rarely positive when headlines refer to fungi, bacteria, algae and viruses (see Figure 10.1). Such publicity has possibly contributed to some of the concern students and teachers have about work with microbes (Lock, 1996); all the more reason for including such work in our teaching in order to redress the balance. Microbes that students might have seen or heard about may include yeast, mould, mildew and rust on plant leaves or diseases linked to microbes, such as meningitis, flu, measles and AIDS. With such a legacy of ideas brought to lessons, the importance of a positive approach is readily apparent. Students may need reassurance that the majority of microbes are not harmful.

The majority of students will have had experience of microbes in their primary schools and might know that there are small organisms called microorganisms, which grow and reproduce and which may be harmful or beneficial. At the end of primary school they may have considered issues such as food decay and preservation (DfES, 2009). However, even with these experiences they are likely to be most familiar, from the media, with terms such as germs, bugs and superbugs (Figure 10.1).

A teaching sequence

■ Using the media

This lesson could be started with a brainstorm to find out what students already know, or it could be preceded by homework that gets them to look at the web, newspapers and television and to report back on what they can find out. A further approach could be to collect pages from papers with cuttings, such as those illustrated in Figure 10.1, and to set an information-finding activity.

The outcomes from such activities could lead to a display which illustrates the balance between positive and negative associations, the use of common terms such as bug, germ and their scientific equivalents and the knowledge that microbes (a convenient shorthand term for 'microorganisms') are a diverse group, including:

- algae (e.g. *Chlorella*)
- fungi (e.g. yeast, mould)
- bacteria (e.g. *Escherichia coli*)
- protozoa (e.g. the malaria parasite)
- viruses (e.g. influenza virus).

The production of such a list enables elements that students already know or have heard about to be drawn together and some tricky issues identified, e.g. where do AIDS (caused by the retrovirus HIV) and CJD (caused by a prion) fit? The development of a spidergram (or concept map) could also achieve the same purpose of helping to organise the information they already know into the scientific diversity of the group.

Figure 10.1
Microbes in the
media

■ Classification

If the class has previously studied the classification of living things using the five-kingdom system then it would be appropriate to remind them that microbes are represented in more than one kingdom and that some kingdoms are exclusively microbes!

Table 10.1 Microbes classified

Types of microbe	Kingdom
bacteria and blue-green 'algae'	monera
protozoa and algae	protoctists
fungi	fungi (not a subdivision of the plants)

Now might be the time to raise the issue of viruses/retroviruses and whether they are living things, although some teachers may prefer to avoid such complications with younger students.

■ Microbe gardens

Microbe gardens can be a popular feature of work at this stage and can provide a convenient link to what microbes look like. Check first that students have not done this activity in their primary school. Algae on bark, macroscopic fungi (mushrooms from the supermarket – best to avoid potentially poisonous ones collected in the field), a block of yeast or yeast tablets, Quorn, mouldy bread and rotting fruit can be used in a small exhibition or circus of activities. The bread and rotting fruit should be in a closed glass or plastic container, a sealed plastic bag would do. The container should be small enough to permit autoclaving without opening before disposal. The reason for enclosure is that it is not known which microbes are growing, although it is unlikely they would be pathogenic (disease causing). In addition, release of spores could trigger an allergic reaction in sensitive individuals if they inhaled enough of them. The bread and fruit should be autoclaved before disposal (see Equipment notes).

 The plastic bag needs to be autoclavable.

■ Misconceptions

The notion that microbes are not living and a means of addressing this misconception has already been mentioned (Choosing a route). Misconceptions linked to the use of the term 'bugs' are common and lead to associations with invertebrate organisms such as insects, other arthropods and other invertebrate animals. Work on classification of living things should help here. As with 'bugs' there are other issues linked to the use of language both in and outside the classroom. Terms such as 'bacteria' and 'virus' are used without distinction and can lead to confusion when antibiotics are not prescribed for illnesses, such as colds and flu, caused by viruses. Here there may be a growing familiarity with the winter flu jab and the availability of anti-virals, for example in relation to the swine flu pandemic.

10.2 What do microbes look like?

A teaching sequence

Microbe gardens provide an initial perspective on variety, but what are microbes like down a microscope? Students need the skills of making wet mount slides and using microscopes that magnify up to

×200. It is good to look at algae (use the green stuff that grows on tree bark, usually a species of *Pleurococcus*, then yeast and mould (both fungi) and finally a bacterium (*Bacillus subtilis* is fine). For the mould use a pure culture of *Mucor* or *Rhizopus* and expose the culture to methanal vapour (a filter paper soaked with 40% methanal (formalin) solution in the Petri dish) for 24 hours to kill the culture and the spores. To protect against allergic reaction to spores, which are killed but can still be released, samples of mould could be removed in the fume cupboard. For convenience, slides of mould could be made up beforehand. With the bacterium a single drop from a broth culture diluted in water on a slide is adequate. Used slides should be collected in a beaker of freshly prepared 1% sodium chlorate(I) (hypochlorite) solution and left for at least 15 minutes. Kill the bacteria before viewing by the addition of methanal solution.

Methanal solution is toxic and corrosive. Use gloves, eye protection and a fume cupboard.

■ Drawings on acetate squares

A good approach is to challenge students to make up slides and then to draw what they can see. They can do their drawings on small squares of acetate (say 10 cm × 10 cm) using overhead projector (OHP) pens, as this provides a convenient way of subsequently rapidly sharing everybody's observations using an OHP, although if your laboratory has a visualiser, scanner or digital camera there may be other, usually slower, ways of achieving the same end with drawings done on paper. If students are working in groups of three then all can make a slide, each with a different organism. However, if you prefer to do things sequentially, start with algae, then yeast and moulds, leaving the bacteria till last or for the fastest workers. Helping students to locate areas of the slide to observe and guiding them to make drawings of what you are hoping they will see is an essential teacher pedagogy in this part of the lesson.

When sharing observations, start with the algae then yeast because these are biggest and most students will have made relevant observations. With the algae students may draw isolated cells, groups of two, four and big groups (there will also be other sized groups). A good question to ask students is, 'How can we explain one, two, four?'. This links to asexual reproduction where organisms successively divide into two. Students should now be able to explain the threes! One of a pair has divided while the other has not, or both have divided and one has separated. Sharing the drawings ensures that all stages are found. If students have been taught about cell structure in plants, then in algae many common features can be quite clearly seen, e.g. cell walls and chloroplasts. With yeast they see mostly individual cells, but, depending on the species you use, they could see budding or binary fission. The mould will show spores (little round things are common to all the microbes observed so far) and threads from the cotton wool-like bits (mycelium/hyphae).

'Hands up those who've drawn a bacterium. What, nobody? Well, I definitely put some in there, so how do we explain this?' This is a great way into the idea that microbes are different sizes and that these bacteria are so small that students can't even see them down their microscopes. A demonstration with a more powerful microscope and perhaps with a computer-linked camera will show small rod-shaped objects. Yoghurt smears with staining also work well.

■ Size concepts

Students find it difficult to understand the relative size of microbes and a variety of approaches may help. Using microscopes to look at familiar objects, such as lines on graph paper or a hair, can help. A chart that shows the 'p' on a one pence piece magnified ×40 and ×200 (depending on the objective lenses of the microscopes that you have) and the relative size of a yeast cell is good too.

An idea used by a school in Oxfordshire is shown in Table 10.2. Each statement in the table was accompanied by a picture showing the object in question and it formed a frieze that went halfway round the lab. Another approach is to produce a model that students are familiar with and to show microbes on a relative scale, for example, a model pin with a head 20 cm across with onion seeds (representing microbes) stuck on to it.

Table 10.2 An approach to the concept of size

The following was taken from a laboratory wall in Gosford Hill School, Kidlington.	
Anna is holding a metre rule.	Anna's hand is 10 centimetres across (0.1 metres).
Her garden is 10 metres wide.	One of her fingers is 1 centimetre across (0.01 metres).
She lives 100 metres from the bottom of her street.	Each of her eyelids is 1 millimetre thick (0.001 metres).
It's 1 kilometre to the Tesco Superstore.	Each of her eyelids is 1 millimetre thick (0.001 metres).
It's 10 kilometres across her town.	Hairs on her head are 0.0001 metres thick.
Her town is 100 kilometres from London.	Her red blood cells are 0.000 001 metres thick.
Her country is 1000 kilometres long.	Some bacteria are 0.000 000 1 metres long.
The axis of the Earth is more than 10 000 kilometres long.	Small bacteria are 0.000 000 01 metres thick.
100 000 kilometres is a quarter of the distance to the Moon.	Some viruses are 0.000 000 001 metres across.

10.3 Where do you find microbes?

The simple answer is everywhere, with some microbes resistant to high temperatures, long periods of drought and a wide range of pH; some bacteria even live in jet fuel, geysers and rocks. Experimental work with 11 to 13 year olds can involve exposing agar in Petri dishes (plates) to the air in labs and outside for varying lengths of time (30 seconds to 1 hour), or inoculating with water from different sources (tap, river, pond, bottled), soil solutions, finger dabs from washed and unwashed hands, hair, leaves, etc. Don't forget to keep unexposed plates as 'controls' and these can be used to explain sterile technique, ideally accompanied by a teacher demonstration of plate pouring.

The general rule here is not to set out to culture potential pathogens by, for example, culturing from raw meat or using nasal swabs.

■ Incubating plates

If the following lesson is 2 or 3 days later, just keeping the plates in a warm place (away from students) will allow colonies to develop. Plates should be incubated upside down to prevent water dripping on the cultures. If your school has an incubator, then the upper limit for incubation should be 25°C, maximum 30°C, in order to avoid selecting for organisms that are adapted to human body temperature (ASE, 2001). However, yoghurt cultures can be incubated above 30°C.

A teaching sequence

■ Pouring plates

Students should understand how agar plates are produced. Start by showing students Petri dishes (Julius Petri – see Enhancement ideas) as supplied by the manufacturer in sealed packs that have been irradiated with gamma radiation, say, a cobalt-60 source, which kills all microbes and leaves the dishes sterile. You can demonstrate the pouring of plates (see 'Where do you find microbes?') and/or students can pour their own. The plates poured by students should be taped (see Figure 10.2) until it is clear that the agar is still sterile (i.e. no colonies develop within 48 hours). The sterile plates can then be used for inoculation. Laboratories should be draft-free to minimise lateral mass flows of air and all work should be carried out close to a Bunsen burner on a non-luminous flame where air, and any spores suspended in the air, will be moving upwards.

Wash hands before and after the demonstration.

■ Inoculating plates

Students can practise inoculation of plates using yeast suspensions, wire loops and spreaders, but plates should be taped following inoculation and remain unopened unless the cultures have been killed. A more detailed consideration of safety can be found in ASE (2001) *Topics in Safety* (3rd edition). Elements of safety specific to teachers are included in the Equipment notes section on page 307.

This activity can be turned into a simple investigation by making agars with different concentrations of sugar and seeing what impact this has on the growth rate of yeast. Inoculating yeast into wells cut into the agar with cork borers and measuring the surface area of the yeast colony is an effective way of achieving this though plates should not, on this occasion, be incubated upside down. This can provide a good activity for supporting students in obtaining sufficient valid and reliable data; an appropriate link to key processes in How Science Works.

■ Taping and labelling plates

Any plates used in investigations such as those mentioned in the previous sections, should be taped and labelled as shown in Figure 10.2. Ideally, label the base rather than the lid (in case the lid gets separated from the base) and do this just before the inoculation is set up, rather than afterwards. Some textbooks show tape going right across the base or lid of the Petri dish. This strategy should not be followed as discoloured tape could prevent quality observation at a later point. Worse still, some suggest taping around the rim of the plate. Petri dishes are designed with, usually, three small lips inside the rim of the lid which raises the lid above the base. This permits diffusion of gases into/out of the dish (but spores remain inside). This explains why some cultures smell when they are growing, but remember that the presence of a smell does *not* indicate that there had been contamination or escape of spores. Taping around the rim would lead to the oxygen inside the plate being depleted and could encourage organisms that grow without oxygen (anaerobically) to multiply, many of which are pathogens.

Figure 10.2 Taping
and labelling Petri
dishes

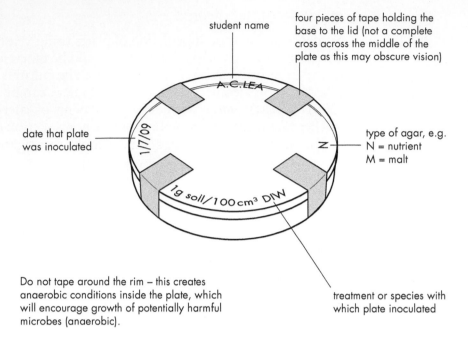

student name

four pieces of tape holding the
base to the lid (not a complete
cross across the middle of the
plate as this may obscure vision)

A.C.LEA

date that plate
was inoculated

1/7/09

type of agar, e.g.
N = nutrient
M = malt

N

1g soil/100 cm³ DIW

Do not tape around the rim – this creates
anaerobic conditions inside the plate, which
will encourage growth of potentially harmful
microbes (anaerobic).

treatment or species with
which plate inoculated

■ Macroscopic differences between fungi and bacteria

By using malt and nutrient agars it is possible to investigate what type of microbes grow best on each type of agar. To do this investigation students need to recognise the differences between fungal and bacterial colonies. The first step is to appreciate that a circular colony has developed from a single cell/spore landing on the agar. This introduces the issue of reproduction too. A single bacterium, under ideal conditions, can divide into two once every 20 minutes. Students could be encouraged to estimate how many cells there could be in a colony that is 24 hours old. Calculators and/or a computer spreadsheet can be used to demonstrate the rate of growth. The main differences between bacterial and fungal colonies are shown in Table 10.3, although not all features are shown by all colonies. Examination of colonies, using hand lenses, is possible through the lids of Petri dishes. Students should easily see the mycelium and spore containers (sporangia) in fungal colonies.

Table 10.3 Macroscopic differences between fungi and bacteria

	Fungi	Bacteria
Colony size	Large	Smaller
(seed from above)	may fill the whole dish	1 mm to 10 mm diameter
Colony profile	Tall	Falt
Appearance of colony surface	Surface is dull	Surface is shiny
Texture	Like cottonwool	Like a drop of liquid
Colour	Grey, white (spores are sometimes coloured, e.g. black, blue–green)	Yellow, pink, white, a range of colours
Growth medium	Grow best on malt agar (more alkaline)	Grow best on nutrient agar (more acidic)

Now students can estimate how many microbes sediment from the air on to a Petri dish in 1 minute, how many live in 1 gram of soil or on the pad of one washed finger. This whole area lends itself to a wide range of investigations and questions that students can explore. Some ideas are suggested in the Further activities section.

10.4 How do you handle microbes safely?

■ Assessing risk and working safely

At both Key Stages 3 and 4 in England students should be able to 'assess risk and work safely in the laboratory' (QCA, 2010, p. 209), a key element of the skills and processes that comprise part of How Science Works.

One way of working towards this objective is to get students involved in developing their own safety rules and an example of these is illustrated in Figure 10.3. Most of these will have links to food hygiene in the home, links which should be made explicit as, for example, in items 1 and 2 of the pupil rules for working safely with microbes.

In Scotland, refer to the SSERC website.

10.5 What do microbes need to grow?

A teaching sequence

■ Broth experiment

Work in this section can provide opportunities to introduce important work carried out on microbes by Pasteur. The diagram in Figure 10.4 shows an experiment with nutrient broth in the test tubes; the cloudiness of the broth is an indicator of the extent of bacterial growth. Students can set up these experiments. Tubes 1 and 3 (and others if students' sterile technique is poor) would need 1 cm^3 of 40% methanal solution added before handing the tubes back to students to examine. Tube 2 should illustrate that microbes are destroyed by boiling and that 'spontaneous generation' of living organisms does not occur. This element alone may be addressing an important misconception about microbes. Tube 4 replicates the experiments that Pasteur did with swan-necked flasks (see Further activities on page 306) and has clear links to the shape of air locks used in home brewing activities. Tubes 1 and 3 show that bacterial contamination is by direct sedimentation of spores from the atmosphere, which relates back nicely to the pouring and inoculating plates activities, as well as having clear links to food hygiene.

Ensure the safety rules in Figure 10.3 are followed. Seal tubes 1 and 3 to prevent spilling of methanol.

Figure 10.3
Student rules for
working safely with
microbes. (After an
original idea by
Lindsay Hudson.)

ALL MICROBES CAN BE DANGEROUS IF NOT TREATED CORRECTLY

1 Always wash your hands before starting any experiment in microbiology, and again before leaving the lab.

2 Cover exposed cuts with waterproof dressing.

3 As soon as the plates have been used, each lid must be attached to the base with adhesive tape, which must NEVER be removed.

4 If a culture is accidentally spilled, report it to a teacher AT ONCE so it can be cleaned up immediately and with complete safety.

5 After you have finished with a culture, dispose of it AS DIRECTED – NEVER TAKE IT HOME.

6 Don't lick your hands; don't eat in the lab.

7 Wash your hands before leaving the lab or immediately afterwards.

DO ALL THIS AND MICROBES WILL BE FUN.

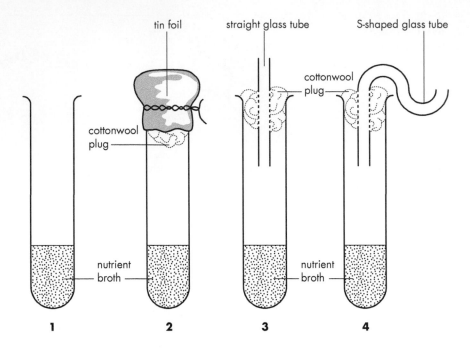

Figure 10.4 Broth experiment. What do microbes need to grow?

■ Food preservation

In the experiment in the previous section a qualitative measure is used to estimate microbial growth – the turbidity of the broth solution. In a development of this investigation students can compare the effect of different preservatives on foodstuffs by estimating the degree of decay of peas through observing the turbidity of a microbe suspension, although colorimetry could also be used (Practical Biology, 2010).

There is a range of other experiments where factors influencing growth or growth rates of microbes can be explored. Growth can be measured by colony diameter (for fungi) or directly by counting cells/spores on a micrometer slide, although these methods are mainly appropriate for 16 to 19 year olds. With an organism like yeast, indirect measures of activity can be made by measuring the evolution of carbon dioxide. Temperature is an obvious factor to investigate with 11–13 year olds and one which lends itself well to incorporating elements of How Science Works. Students could investigate whether refrigeration slows microbe growth but does not stop it; whether freezing and boiling both stop growth and whether the microbes start to grow again when removed from the freezer and when the boiling water has cooled down. Table 10.4 shows a range of factors that influence microbe growth and illustrates how this can be used to our advantage in food production and preservation. Such work makes a clear link with the involvement of microbes in a healthy life.

Table 10.4 Microbial growth and food preservation

Method	How bacterial growth is affected	Example
Ultra-heat treatment (UHT) (135 °C for 2 seconds)	Microbes and resistant spores killed	UHT Longlife milk
Sterilised by boiling (100–115 °C for 15 minutes)	Microbes killed	Sterilised milk
Pasteurised (70 °C for about 15 seconds)	Many microbes killed	Pasteurised milk
Canning (boiling and sealing to exclude oxygen)	Microbes killed and no oxygen available	Tinned beans
Sterilise by irradiation (e.g. from cobalt 60 source)	Microbes killed by gamma radiation	Salmon
Freeze drying	No water for microbial growth	Instant potato
Preserving by adding sugar or salt	Microbes plasmolysed, water moves out of cells	Jams (sugar), fish (salt)
Pickling	pH too acidic for microbes to grow	Pickled onions
Fridge (4 °C)	Microbe growth slowed down	Yoghurt
Freezer (-18 °C or lower)	Microbe growth stopped	Frozen vegetables

10.6 How are microbes involved in a healthy life?

Headlines and news stories about 'bugs' that affect humans, such as those listed in Figure 10.1, intrigue and interest students and therefore provide an excellent starting point for work on microbes and health, as well as a stimulus for individual investigative projects by older students. These headlines also provide a context for developing an understanding about fundamental microbiological principles and the ways in which microbes affect the lives of all animals and plants. Earlier work on the topic will have highlighted the need for hygienic laboratory practice and students will be aware of the effect of factors such as temperature on the growth of microbes.

A teaching sequence

■ Hygiene and food preparation

Despite greater public awareness about the need for food hygiene, and legislation about the storage and preparation of food in commercial organisations and industry, outbreaks of food poisoning are still common – and most are preventable. Outbreaks of food poisoning are usually caused by people handling food with dirty hands, insufficient cooking or storing food incorrectly, e.g. at the wrong temperature or leaving food uncovered. Students could be asked to find out the reasons for the advice on food hygiene given in Table 10.5.

Table 10.5 Advice on food hygiene (BUPA, 2010)

● Rinse your hands well and dry them on a clean hand towel, a disposable paper towel, or under a hand dryer.
● Use different cloths for different jobs (e.g. washing up and cleaning surfaces). Wash them regularly on the hot cycle or soak in a dilute solution of bleach.
● Don't handle food if you have stomach problems such as diarrhoea and vomiting, or if you're sneezing or coughing frequently.
● Cover up cuts and sores with waterproof plasters.
● Use separate chopping boards for preparing raw meat, poultry and seafood and for fresh produce such as salads, fruit and vegetables.
● Store fresh or frozen food in the fridge or freezer within 2 hours of purchase – sooner if the weather is hot.
● Keep the fridge at less than 5 °C and the freezer at less than −18 °C – consider getting a thermometer.
● Follow the recipe or packet instructions for cooking time and temperature, ensuring the oven is pre-heated properly.
● When microwaving, stir food well from time to time to ensure even cooking.
● Only reheat food once and serve piping hot.
● Don't serve eggs with runny yolks, or egg-containing foods that won't be cooked, for example homemade mayonnaise.

Reasons that students might give should show their awareness that bacteria can spread from raw food, in particular meat, to food that has already been cooked or to food that is eaten raw, such as salads. In addition they should understand the importance of storing food in the right place (e.g. fridge or freezer) and at the correct temperature. They should know too that if food isn't cooked at a high enough temperature, bacteria can still survive and that eggs could contain harmful bacteria, which can be dangerous to pregnant women, older people and babies.

Students can also investigate different packaged foods to work out how long they can be stored and what 'sell by' and 'use by' dates mean. These investigations can be linked to work on food preservation.

■ Investigation of the effect of hand washes on bacteria

Advertisers make much of the ability of disinfectants/hand washes and toilet bowl cleaners to 'kill all known germs'. Students can investigate the truth of such claims in a number of ways and these investigations provide good links to How Science Works. It is possible to provide students with a range of anti-bacterial hand washes, some famous brands and some supermarket 'own' brands, as well as an alcohol hand wash of the sort used in hospitals. Students can be asked to design their own investigations and to control as many factors as possible. They can consider how they will standardise their techniques, which, if finger dabs on agar plates are being used, might include washing and drying their hands in the same way and using the same volume of water and hand wash. Students can be told that this is close to the standardised method for hand washing used in the soap industry. For more details from a teacher who used a technique very similar to this see Thomas (2010). This practical provides links to the early work of Semmelweis, a Hungarian who investigated the impact of surgeons washing their hands on the spread of infection in hospitals.

An alternative approach that does not involve pressing fingers onto agar plates is to use plates that have been inoculated with lawn cultures of safe microbes. Filter paper discs impregnated with domestic toiletries such as toothpastes, deodorants or mouthwashes are then pressed onto the lawn culture and the size of the halo, if any, of lysed bacteria around the disc after, say, 48 hours can be measured. As some of the products such as toothpaste and mouthwash may be diluted in use it is appropriate for students to investigate the effect that this might have on their efficacy.

Students are typically highly motivated by work that relates to bodily functions and fluids, and there is an engaging investigation on 'How good is your toilet paper?' that can be found on a practical biology website complete with relevant risk analysis (Practical Biology, 2010) (www.practicalbiology.org). This practical work provides further opportunities for involving processes and skills linked to How Science Works.

■ Clean water

What do students think is meant by 'clean' water? Water from a spring, a well or a tap, that looks 'clean', may contain disease-causing microbes. Untreated water supplies are still a major source of ill-health, infant mortality and epidemics in many parts of the world. The story of how the spread of cholera was halted in London in the middle of the nineteenth century as a result of careful 'detective' work can provide a starting point for students to investigate other disease-causing microbes. There are also opportunities for drama as well as for data collection and analysis allied to investigative journalism, where students prepare questions (and answers) for experts. Students might use the internet to find out more about pressure groups concerned about water, e.g. Surfers Against Sewage.

The presence of bacteria in water can be shown by inoculating agar plates with samples of water from different sources as described earlier in this chapter. Water can also be drawn through filters that are then stained with a dye to show the presence of bacteria. Students may need reassurance that the large numbers of bacteria demonstrated by this method are not all 'harmful' to humans. Students can find out about the principles of the first stages of water treatment and purification by filtering muddy water samples: this activity can be done as a problem-solving activity. Chemical purification of filtered samples can be demonstrated and the treated samples tested for bacterial content. Information and posters about water treatment are available from water companies; students can also create their own posters and provide advice for consumers on the importance of 'clean' water. Good hygiene is essential when handling muddy water.

■ Digestion

Microbes play an important role in digestion in many animals. They are particularly important in the digestion of plant material, such as cellulose and wood, which is not readily broken down in the guts of animals. The guts of herbivorous mammals, such as cows and sheep, contain very large numbers of bacteria that produce cellulase, an enzyme that breaks down (digests) cellulose and that mammals cannot produce themselves. The simpler molecules produced after the breakdown of cellulose by cellulase can be metabolised (used) by the bacteria, which subsequently release chemicals into the gut of the mammal that can be absorbed and used.

Microbes also play a key role in the breakdown of waste material produced by animals and plants that is linked to the nitrogen cycle. Students can observe how microbes break down, i.e. 'digest', plant material by looking at compost heaps or piles of lawn clippings.

■ Microbes and disease

Section 10.1 introduced students to the range of different microbes, including examples of those that cause diseases in humans, such as colds and influenza. Microbes cause disease in plants and other animals – mice and rats, as well as humans, can suffer from malaria! Research into the history and epidemiology of diseases, such as bubonic plague, smallpox, influenza and cholera, can be used to provide insights into how widespread diseases are and how they affect populations in different parts of the world. Students can be encouraged to report their findings in a variety of ways, including drama or the use of databases and spreadsheets. This work leads naturally into a study of the development of vaccines and, for older students, a study of the functioning of the immune system.

■ Antibiotics and antibodies

These are terms that are frequently confused by students. Most students will have been prescribed antibiotics at some time in their lives; however, many will be confused about what sorts of infections are treatable by antibiotics. A brainstorm provides a good starting point to identify what antibiotics are and what they can do. Students could undertake library research to find out the causes of different types of infection, such as viral, bacterial and fungal, and how these are treated. Follow-up work can include the story of the development of penicillin, Alexander Fleming and practical demonstrations to illustrate the effect of antibiotics on bacterial growth. It is possible to buy discs impregnated with different antibiotics at different concentrations. These can be placed on lawn cultures of microbes so that students can find out for themselves that antibiotics, such as penicillin, are medicines that help to cure bacterial disease by killing infective bacteria. This experiment can also show that it is important for specific bacteria to be treated by specific antibiotics as not all antibiotics kill all bacteria.

Students should know that the use of antibiotics has greatly reduced deaths from infectious bacterial diseases but that overuse and inappropriate use of antibiotics, such as using them to treat viral infections, has increased the rate of development of strains of bacteria that are antibiotic resistant. Some bacteria, such as MRSA, have developed resistant strains as a result of naturally occurring mutations in bacterial populations on which natural (or unintentional artificial) selection then acts.

Viruses are pathogens which mainly exist and reproduce inside the cells of other organisms such as animals or plants. For this reason it is difficult to produce a drug that destroys the virus without also killing the body cells.

It is possible to immunise against a disease by introducing small quantities of dead or inactive forms of a pathogen into the body, a process called vaccination or immunisation. Vaccines stimulate white blood cells to produce antibodies – chemicals which destroy pathogens. This may make the person immune to future infections by the pathogen because the body can respond by rapidly making the correct antibody, in the same way as if the person had previously had the disease.

10.7 How are microbes involved in biotechnology industries?

Microbes are involved in the production of many different products, including:

- food and alcohol
- medicines and drugs, e.g. antibiotics, insulin
- fuels (e.g. gasohol)
- transgenic organisms.

A teaching sequence

■ Microbes in food and other industries

The importance of microbes in industrial processes is best introduced by starting with familiar food products, such as yoghurt, that can be made readily in the laboratory. Other activities could include the making of bread or ginger beer and the fermentation of fruit juices to produce alcohol. Sensors linked to a computer can be used to show the rate of microbial activity. For example, a position sensor can show how bread dough rises – starting off slowly, rapidly rising and then slowing down. Activities of this type can be undertaken by students of all ages. Students will need to be reminded not to taste any of the products made in the laboratory, unless it is suitable for Food technology.

The National Centre for Biotechnology Education at Reading University (NCBE, 2010) provides more than 60 practical protocols using microbes, which are all freely available from their website. Many of these involve scaled down industrial applications that can be carried out safely in school laboratories, some in a single science lesson.

Microbes are important in a range of industrial processes, including the pharmaceutical industry. Some of these processes can be replicated by older students, e.g:

- the production of plastic from bacteria
- amylase production by yeast
- the breakdown of cellulose waste from plants by bacteria.

■ Microbes and genetic engineering

The implications of introducing genetically modified food plants, for example tomatoes, soya, maize and potatoes, into the human food chain have been the subject of considerable debate and public concern in some countries, including the UK. The basic principles of genetic engineering are relatively easy to understand if students have some understanding of cell structure and DNA. For this reason this topic is better suited to students from 14 years upwards. Physical models can be useful in demonstrating how sections of plasmids in bacteria can be modified to incorporate new genetic material into the cells of animals and plants. Students can extract DNA from cells (NCBE, 2010). Examples of genetic engineering include the production of insulin for use by diabetics from genetically engineered bacteria or yeast (Reiss & Straughan, 1996; Kjeldsen & Andersen, 1997).

■ Ethical and moral issues

The advantages of including work on topics such as genetically engineered organisms is that it provides an ideal opportunity to give students experience of examining the ethical and moral implications of using and applying science. One of the key reasons for teaching science to all students in our schools up to the age of 16 years is that as fully participating members of a democratic society, we need to be engaged in considering the balance between the advantages and disadvantages of developments offered by science before making irrevocable decisions concerning them, for example issues linked to the selective breeding, genetic engineering and cloning of animals and plants. Although debates, discussions and role plays are not currently strong pedagogies in the repertoire of science teachers, growing use is being made of them and there are alternative approaches where students research issues using the internet and feed back to their colleagues, possibly using a writing/reporting frame provided by their teacher.

■ Microbiology outside the classroom

Visits to places such as dairies, breweries or pharmaceutical plants can provide insights into a number of microbiological principles and applications, including the conditions needed for the growth of microbes and the need for hygiene practices to ensure that contamination of, for example, foodstuffs does not occur. The size and complexity of the equipment used in the production of products such as beer is a revelation for most students; identifying the problems faced by industrialists in 'scaling up' processes helps to raise issues about sterilisation of equipment as well as the biology of fermentation. Asking students to estimate how many yeast cells are present in a fermentation vat reinforces ideas of size and replication encountered earlier. DVDs may provide a useful alternative where industrial visits are not possible.

Further activities

- Ideas for investigations include the following:
  - How effective are different disinfectants at killing bacteria?
  - How does dilution influence the actions of disinfectants? (You can ask the same question about toothpaste, deodorant, anti-perspirant, antiseptics, Milton, etc.)
  - Is it best to let dinner plates drain, dry them with a tea towel or use a dishwasher? (The surface of plates can be sampled for spores using sticky tape.)
  - Can bacteria get through toilet paper?
  - Which toilet paper gives the best protection against bacterial contamination of the hands?
  - What are live yoghurts?
- Ideas for information seeking and presentation activities using written sources or the internet can involve the role of these scientists in the study of microbes:
  - Louis Pasteur
  - Joseph Lister
  - Alexander Fleming
  - Robert Koch
  - Edward Jenner
  - Ignaz Semmelweis
  - Julius Petri
  - Paul Ehrlich
  - Dmitri Ivanovski.

There are no women in this list but you might like to explore the role that Pasteur's wife played in his research and the observations of Lady Mary Wortley Montagu on inoculation against smallpox.

Enhancement ideas

This is a list of further issues and questions that students can be invited to research, possibly as a homework activity using library resources, e.g. the internet.

- Very few organisms other than microbes can digest cellulose. For this reason most animals feeding on plants are dependent on microbes for digestion.
- Many herbivores have special gut structures where microbes live, e.g. cows with 'four stomachs', rabbits with a caecum.
- Rabbits refaecate; they eat their own faeces. Because the caecum, where the bacteria are, is near the end of the gut, few of the digested nutrients are absorbed into the blood during the first passage of food through the gut. Rabbits produce soft green faeces, usually at night. They eat these and during the second passage of food through the gut absorption is more effective and the familiar dry, roundish faecal pellets are produced.
- Where do young herbivores get their gut flora from?
- Why do some people become constipated when taking antibiotics?
- Why don't doctors prescribe antibiotics for a viral infection?
- Body odours are caused by microbes breeding in the sweat. What is the difference between an anti-perspirant and a deodorant?
- Why do cows produce pats? Is it because digestion of the cellulose is so effective that there is no fibre to hold the faeces together?
- Why is horse dung so full of fibre? Is it because it is not very efficient at digesting cellulose. If so, why isn't it?

Equipment notes

■ Early planning/advance organisation

- Two months before work with microbes you should order safe species from reputable suppliers.
- Three weeks before work, liaise with technicians. Plates and media can be made up and kept in the fridge, but problems with condensation may result.
- Unlike some biology topics, work with microbes can be done at any time of the year.

■ Autoclave/pressure cooker

This is an essential item both for sterilising agars, broth and any other recyclable equipment and for destroying cultures before disposal. Roasting bags are a cheap means of containing plates/cultures during autoclaving and for later disposal with refuse. Autoclaving at $100\,kN\,m^{-2}$ (or 15 lb/square inch) for 15 minutes destroys all microbes and spores.

■ Inoculation chamber

It is not essential to have an inoculation chamber for most microbiology work in school – working in an area close to a Bunsen burner on a non-luminous flame that creates an updraft is adequate. However, transfer chambers are not prohibitively expensive and can be used for a range of other activities such as tissue culturing.

Other resources

Useful organisations

Association for Science Education, College Lane, Hatfield AL10 9AA, and especially www.gettingpractical.org.uk/

CLEAPSS, School Science Service, Brunel University, Uxbridge UB8 3PH; www.cleapss.org.uk/

National Centre for School Biotechnology, School of Chemistry, Food Biosciences and Pharmacy, University of Reading, Reading RG6 2AJ; www.ncbe.reading.ac.uk/

Society for General Microbiology; www.sgm.ac.uk/

MiSAC (Microbiology in Schools Advisory Committee) is supported by the Society for General Microbiology (see above) and their websites include more safety information and a link to ask for advice by email; www.microbiologyonline.org.uk/

Science and Plants in Schools (SAPS). Lots of good practical activities using single-celled algae; www-saps.plantsci.cam.ac.uk/

Science Learning Centres run courses on practical microbiology; www.sciencelearningcentres.org.uk/

SSERC, the Scottish Schools Education Research Centre gives bulletins and resources.

Books

ASE (2001). *Topics in Safety* (3rd edition). Hatfield, Association for Science Education.

ASE (2006). *Safeguards in the School laboratory*. (11th edition). Hatfield, Association for Science Education.

Lock, R. (1993). Use of Living Organisms. In Hull, R. (ed.) *ASE Secondary Science Teachers' Handbook*, pp. 179–205. London, Simon and Schuster.

References

ASE (2001). *Topics in Safety* (3rd edition). Hatfield, Association for Science Education.

BUPA (2010). www.bupa.co.uk. Accessed 6 October 2010.

DfES (2009). www.standards.dfes.gov.uk/schemes2/science/sci6b. Accessed 1 October 2010.

Kjeldsen, T. and Andersen, A.S. (1997). Insulin from yeast, *Biological Sciences Review*, **10** (2), pp. 30–32.

Lock, R. (1996). Educating the 'New Pasteur'. *School Science Review*, **78** (283), pp. 63–72.

NCBE (2010). www.microbiologyonline.org.uk/teachers/resources. Accessed 6 October 2010.

Practical Biology (2010). www.practicalbiology.org/ Website developed by the Nuffield Foundation and Society of Biology.

QCA (2010). www.qca.org.uk/curriculum. Accessed 1 October 2010.

Reiss, M. and Straughan, R. (1996). *Improving Nature? The Science and Genetics of Genetic Engineering.* Cambridge, Cambridge University Press.

Thomas, S. (2010). Involving the soap industry in science lessons. *School Science Review*, **91** (337), 21–24.

Acknowledgement

To Sheila Turner with whom this chapter was originally written and whose influence is strongly reflected in this revised version. Her influence in teaching secondary school biology lessons continues.

Appendix

A book of this sort cannot contain everything. Many readers will want to use it in conjunction with a good student textbook. In addition, the following books, journals organisations and other resources are recommended as valuable sources of information and advice.

Books

Association for Science Education (2006) *Safeguards in the School Laboratory, 11th Edition*, Hatfield, Association for Science Education.

Brown, C. R. (1995) *The Effective Teaching of Biology*, London, Longman.

Campbell, P. (ed.) (2010) *The Language of Measurement: Terminology used in school science investigations*, Hatfield, Association for Science Education.

Darlington, H. and Bell, J. W. (2010) *Biology: Ideas and experiments – Science Notes* from School Science Review, Hatfield, Association for Science Education.

Hollins, M. (ed.) (2011) *ASE Guide to Secondary Science Education, new edn*, Hatfield, Association for Science Education.

Oversby, J. (ed.) (2011) *ASE Guide to Research in Science Education*, Hatfield, Association for Science Education.

Reiss, M. (ed.) (1996) *Living Biology in Schools*, London, Institute of Biology.

Journals

Catalyst www.sep.org.uk/catalyst/

Journal of Biological Education www.societyofbiology.org/education/educational-resources/jbe

School Science Review www.ase.org.uk/journals/school-science-review/

Science Teacher Education www.ase.org.uk/journals/science-teacher-education/

Organisations

Association for Science Education www.ase.org.uk/home/

CLEAPSS www.cleapss.org.uk/

National Centre for Biotechnology Education www.ncbe.reading.ac.uk/

Nuffield Foundation www.nuffieldfoundation.org/

Science and Plants for Schools www.saps.org.uk

Society of Biology www.societyofbiology.org/

Wellcome Trust www.wellcome.ac.uk/

Other resources

Association for the Study of Animal Behaviour educational resources http://asab.nottingham.ac.uk/education/index.php

British Ecological Society educational resources http://www.britishecologicalsociety.org/educational/

National STEM Centre elibrary www.nationalstemcentre.org.uk/elibrary

Practical Biology website www.practicalbiology.org/

Index